Statistical Techniques in Simulation

(IN TWO PARTS)

PART I

STATISTICS

Textbooks and Monographs

VOLUME 1: The Generalized Jackknife Statistic
H. L. Gray and W. R. Schucany

VOLUME 2: Multivariate Analysis
Anant M. Kshirsagar

VOLUME 3: Statistics and Society
Walter T. Federer

VOLUME 4: Multivariate Analysis: A Selected and Abstracted Bibliography, 1957-1972
Kocherlakota Subrahmaniam and Kathleen Subrahmaniam

VOLUME 5: Design of Experiments: A Realistic Approach
Virgil L. Anderson and Robert A. McLean

VOLUME 6: Statistical and Mathematical Aspects of Pollution Problems
edited by John W. Pratt

VOLUME 7: Introduction to Probability and Statistics (in two parts). Part I: Probability. Part II: Statistics
Narayan C. Giri

VOLUME 8: Statistical Theory of the Analysis of Experimental Designs
J. Ogawa

VOLUME 9: Statistical Techniques in Simulation (in two parts)
Jack P. C. Kleijnen

OTHER VOLUMES IN PREPARATION

Statistical Techniques in Simulation

(IN TWO PARTS)

JACK P. C. KLEIJNEN
Katholieke Hogeschool
Tilburg, The Netherlands

PART I

MARCEL DEKKER, INC. New York

MARCEL DEKKER, INC.
270 Madison Avenue, New York, New York 10016

LIBRARY OF CONGRESS CATALOG CARD NUMBER: 74-79920
ISBN: 0-8247-6157-X

Current printing (last digit):
10 9 8 7 6 5 4 3 2

PRINTED IN THE UNITED STATES OF AMERICA

To Mai and Wilma

CONTENTS OF PART I

CONTENTS OF PART II

PREFACE

This book discusses the statistical design and analysis of simulation and Monte Carlo experiments, with emphasis on digital simulation with models of management and economic systems. Simulation is always a sampling experiment whenever the model contains one or more stochastic variables (although it is a special type of sampling experiment since simulation is performed on an abstract model instead of a real-life object). In any statistical experiment a careful design and analysis is desirable. The analysis should extract as much information from the experiment as possible and, moreover, should reveal the limitations of conclusions based on sampling data. The design should ensure that the experiment contains as much information as possible.

Though simulation is a sampling experiment the technique is usually not applied by statisticians but by engineers, social scientists, etc. (since simulation also requires model building). A thorough statistical design and analysis has been lacking in most simulation studies. It is our hope that this book will make the users of simulation more aware of the statistical aspects of simulation. The purpose of our book is to give the reader a working knowledge of the statistical techniques involved. Because so many techniques are relevant we had to restrict attention to a selected number of topics (but Chapter II gives an overall picture of the statistical aspects of simulation). We have further tried to make all chapters independent of one another, at least to a high degree. In this way the practitioner interested in a specific problem (e.g., runlength determination) may read only the relevant part of the book. While we aim at working knowledge of certain techniques, references are given for further study of related topics. (Each chapter ends with a long list of references; many chapters have a literature section which recommends particular references as a guide for additional study. We would appreciate very much if the reader informed us of any relevant literature omitted by us. We especially welcome references to publications that have recently become available.) To read this book only a "basic" knowledge of mathematical probability and statistics is required. We conjecture that most management scientists and economists do have knowledge of dependence of events, t-tests, simple regression analysis, etc.

Before proceeding to a brief survey of the book we point out that each chapter begins with a section that gives a more detailed summary. Chapter I describes systems, models, and Monte Carlo and simulation techniques. It further discusses random number generation and sampling of stochastic variables. Its purpose is not to present new ideas but to define basic concepts. It also contains appendices with simple applications of the Monte Carlo and simulation methods. Chapter II gives a survey of the statistical aspects of simulation. It further shows how the following chapters fit into the total picture. Chapter III presents variance reduction techniques that may have wide applicability in complicated simulation experiments: stratification, selective sampling, control variates, importance sampling, antithetic variates, and common random numbers. This chapter results in extensions, limitations, and corrections of existing techniques. It also contains an algorithm for the optimal allocation of computer time when applying antithetic and common random numbers. Chapter IV presents experimental designs relevant in simulation. A detailed discussion is given of general factorial experiments (and their analysis of variance and regression analysis), 2^{k-p} designs, and random, supersaturated, and group-screening designs. It further contains a bibliography on response surface methodology. Chapter V covers the effect of the sample size on the reliability of statements based on a sample. It consists of three independent parts. Part A discusses the evaluation of a single population or the comparison of only two populations (or systems). The sample size may be fixed or may be so selected that a predetermined reliability results. Part B covers multiple comparison procedures, i.e., the sample sizes are fixed and k (≥ 2) systems are compared with one another. It also discusses various types of error rates (e.g., the experimentwise rate), selection of a subset containing the best population, and efficiency and robustness of the procedures in simulation. Part C gives multiple ranking procedures, i.e., how many observations should be taken from each of the k (≥ 2) systems in order to select the best system. (The "indifference zone" approach is followed.) Again the efficiency and robustness of these procedures are discussed. Finally, Chapter VI presents a case study demonstrating how various techniques of the preceding chapters can be applied. The case study concerns a Monte Carlo experiment with Bechhofer and Blumenthal's multiple ranking procedure, and investigates the robustness of this procedure. All chapters contain a number of simple exercises and references to the literature for additional exercises. (Note further that stochastic variables are underlined in this book to distinguish them from deterministic quantities.)

Jack P. C. Kleijnen

ACKNOWLEDGMENTS

It is a pleasure to thank the many persons who made this publication possible. Since I began working as a research associate at the Katholieke Hogeschool in Tilburg (Tilburg School of Economics and Business Administration) Professor M. Euwe and later on Professor G. Nielen have given me ample time to engage in the research of the statistical aspects of simulation.

The Netherlands Organization for the Advancement of Pure Research (Z.W.O.) and the Board of the Katholieke Hogeschool enabled me to do research at the University of California at Los Angeles during the academic year 1967-1968. The Katholieke Hogeschool and Professor T. Naylor made it possible for me to visit Duke University during the summer of 1969.

My discussions with Professor H. Lombaers (Technische Hogeschool Delft) resulted in many adaptations of the original Chapter I. The following persons gave their useful comments on parts of the original papers on which this book is based: Professor J. Kriens, Professor D. Neeleman, Mr. H. Tilborghs, Mr. A. van Reeken, Dr. C. Weddepohl (all at the Katholieke Hogeschool), Professor R. Nelson and Professor J. MacQueen (University of California), Professor D. Graham and Professor T. Naylor (Duke University), Professor T. Wonnacott (Western Ontario University), Professor K. Gabriel (Hebrew University), Professor E. Dudewicz (University of Rochester), and Professor P.A.W. Lewis (Naval Postgraduate School). It stands to reason that any errors remain my sole responsibility.

The camera-ready copy was typed by Miss Rosemarie Stampfel of Redwood City, California. The proofreading was done while the author was a Postdoctoral Fellow at the IBM Research Laboratory in San Jose, California.

Chapter I

FUNDAMENTALS OF SIMULATION

I.1. INTRODUCTION AND SUMMARY

Since this study concentrates on experiments with digital simulation models of business and economic systems we shall first discuss the concepts of system, model, Monte Carlo, and simulation. A discussion of these concepts seems necessary as there is no uniform terminology in the literature [see Naylor et al. (1967a, pp. 2-3)]. It is not our purpose to give rigid definitions, but rather to make clear what we mean when using terms like simulation, Monte Carlo, etc.

Besides terminology we shall give some examples of the simulation and Monte Carlo techniques. At the end of this chapter three appendices follow in which we have worked out simple applications of the Monte Carlo and simulation method. These applications will again be used in the following chapters. In Section I.6.4 we shall discuss random number generation and sampling of stochastic variables, i.e., the basic inputs of a simulation program.

Those readers familiar with simulation may wish to skip this chapter completely except for the glossary at the end (and, possibly, the appendices since we shall refer to them in later chapters; they should also note the recent references on random numbers and sampling of variates).

I.2. SYSTEMS

Following Nelson (1966, p. 1) and van Dixhoorn and Lyesen (1968, p. 15) we describe a system as follows [see also Mihram (1972, pp. 213-222)]: The system is a set of <u>elements</u>, also called components.

For instance, an industrial plant can be looked upon as a system with machines and people as elements, and a national economy can be seen as a system with producers and consumers as components. The elements have certain characteristics or "attributes" and these attributes have numerical or logical values. In the above plant a characteristic of a machine may be whether it is busy or not; in the national economy an attribute of a consumer may be the size of his demand for a particular product. Among the elements relationships exist and consequently the elements are interacting. For example, if there is no operator then the machine can give no service. Besides these so-called internal relationships there are external relationships. The external relationships connect the elements of the system with the environment, i.e., the world outside the system. For instance, there is a certain pattern of customer arrivals into the system. We can represent a system by diagram as in Fig. 1. In Fig. 1 the system itself is represented by the square or "box." Generally speaking the system is influenced by the environment through the input it receives from the outside world. This input is transformed by the process operating in the system. This transformation or process yields the output of the system.

In regard to the dynamic behavior of systems it is necessary to consider the concepts of states and stability. The state of a system is determined by the numerical or logical values of the attributes of the system elements. For instance, if in the above industrial plant the length of the waiting line changes then the system state changes; if in the above economy a consumer changes his preferences then the system moves to a new state. We say that the system is in the steady-state if the probability of being in some state does not

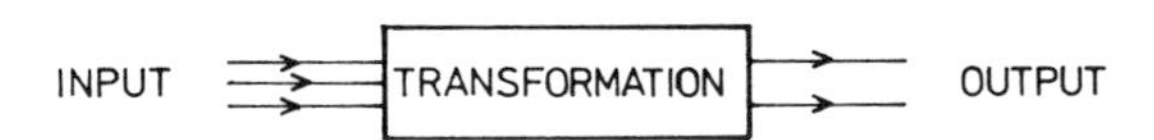

Fig. 1. Graphical representation of a system.

vary over time. Notice that if a system is in the steady state then the system can very well move from one state to another. So there is still action in the system, but the probability distribution of moving from one state to another is fixed. (In the theory of Markovian processes this fixed distribution is called the invariant or stationary distribution.) If the system is not in the steady state then it is in the transient state. The fixed probabilities that hold for the steady state are the limits of the probabilities that are realized after a long period of time. These limiting probabilities are supposed to be independent of the state in which the system started. A system is called stable if it returns to the steady-state after an external shock on the system. Such a shock may be a temporary increase of the rate at which customers arrive in the system. See Conway (1963, pp. 48-49), Kiviat (1967, p. 17), and Nelson (1966, p. 2). Note that in Emery's (1969) collection of papers on the systems approach, the steady state is defined differently since this publication concentrates on deterministic systems.

We refer to the literature[1)] for a discussion of the system concept and classifications of systems such as real-world vs abstract, black box vs identified, open vs closed, adaptive vs nonadaptive, feedback, feedforward and noncontrolled, static vs dynamic, stochastic vs deterministic, continuous discrete and hybrid, technical organizational and abstract systems. Emery (1969) and Rose (1970) have provided some interesting articles on the systems approach to problems in various disciplines, e.g., organization theory, sociology, and biology.

I.3. MODELS

It is possible to define a model very generally as anything that represents something else. Such a definition would cover statues as models of particular human beings, plays representing historic events, etc. Obviously there are many types of models. Churchman et al. (1959, pp. 155-162) and Kiviat (1967, p. 2) discuss iconic, symbolic, and analogous models. Forrester (1961, pp. 49-52) gives an elaborate classification of models. Mihram (1971, pp. 1-17) discusses models and their relationships with simulation. Our study, however, is concerned with what we call abstract models, i.e., models consisting of mathematical symbols or flowcharts. Notice that if the model consists of a single equation, say $z = x + y + c$, then the mathematical formula and the flowchart are essentially the same. However, if we want a model for a complicated system it may be easier to formulate the model as a flow chart rather than as a set of equations and inequalities. An example is the production-inventory model in Naylor et al. (1967a, p. 174) where a flowchart shows the relations among the components of the system. From now on we mean these abstract models when using the word model (unless we explicitly define models differently as we shall do briefly in Section I.6.1).

The object represented by the model is a system. As van Dixhoorn and Lyesen (1968, p. 42) remark the model of a system is an (abstract) system itself. Yet, when in this book we use the word system we want to emphasize that there are elements interacting with each other and with the environment. The word model emphasizes that we are not studying the object of interest but only a representation of it. This representation is assumed to behave essentially in the same way as the object itself.

In economics it is customary to classify the variables in a model as exogenous or endogenous. The values of the exogenous variables are not determined by the model. They can also be called the independent variables. In the terminology of systems theory the independent variables can further be classified as uncontrollable and controllable variables. The uncontrollable variables, e.g., foreign demand, are the inputs into the system. The controllable variables, e.g., government expenditure, are the instrument variables manipulated by certain components of the system. The values of the endogenous variables do depend on the model. In systems terminology we can classify these dependent variables as status or intermediate variables which describe the state of the system, and output variables. Besides the variables of the model we distinguish parameters. Parameters are quantities that influence the endogenous variables, but unlike the exogenous variables they are not variable but constant. (In Section I.6.3 we shall specify "how long" the parameters remain constant in a simulation model.) Relationships describe how the variables and parameters are connected with each other. These relationships can be classified as identities or definitions and operating characteristics. Moreover, we may study several models which differ from each other in so far as they have different exogenous variables, parameters, and relationships. If an exogenous variable, a parameter or a relationship is varied from one variant of the model to another, then it is a factor in the sense of the theory on the statistical design of experiments. If we change one or more factors then the output may also change. In experimental design terminology the output is called the response. See Emshoff and Sisson (1971, p. 52), Hanken and Buijs (1971, pp. 10-11), and Naylor et al. (1967a, pp. 10-15, 322).

I.4. SOLUTION METHODS

In order to answer questions about the object represented by the model we need to "solve" the model. There are analytical and numerical solution methods. An analytical solution uses properties known from that part of mathematics called "analysis" which comprises differential and integral calculus. It gives a solution in the form of a formula that holds for various possible values of the independent variables and parameters. Analytical solutions for simple inventory and queuing problems can be found in Churchman et al. (1959), while Theil and Boot (1962) consider the analytical solution of econometric models consisting of a set of simultaneous linear difference equations of nth order.

A numerical solution substitutes numbers for the independent variables and parameters of the model and manipulates these numbers. Many numerical techniques are iterative, i.e., each step in the solution gives a better solution using the results from previous steps; examples are linear programming and Newton's method for approximating the roots of an equation. Two special numerical techniques are the

Monte Carlo method and simulation. We shall discuss these methods in more detail in the following sections. Notice that the various methods can be combined for the solution of a complicated model.

I.5. MONTE CARLO

Closely following Hammersley and Handscomb (1964, p. 2) we define the Monte Carlo method in a wide sense as any technique for the solution of a model using random numbers or pseudorandom numbers. We shall consider these random numbers in more detail.

Random numbers are stochastic variables which are uniformly distributed on the interval $[0,1]$ and show (stochastic) independence. (Stochastic variables or variates are probabilistic variables; they are underlined to distinguish them from the values they may assume. Random variables are (independent) uniform stochastic variables.) Random numbers have the following two characteristics:

(i) If the $\underline{r}_i$ $(i = 1, 2, 3, \ldots)$ are random numbers then their cumulative distribution, say F, satisfies relation (1) for all values of i (see Fig. 2).

$$\begin{aligned} F(r_i) &= P(\underline{r}_i < r_i) \\ &= r_i \quad \text{for } 0 \leq r_i \leq 1 \\ &= 0 \quad \text{for } r_i < 0 \\ &= 1 \quad \text{for } r_i > 1 \end{aligned} \tag{1}$$

Note that theoretically the random numbers are supposed to be continuous variables with the density function, say f, given in (2).

$$\begin{aligned} f(r_i) &= 1 \quad \text{for } 0 \leq r_i \leq 1 \\ &= 0 \quad \text{elsewhere} \end{aligned} \tag{2}$$

In practice, however, we do not have continuous but only discrete measurements. Suppose we measure with an accuracy of only n decimal digits. Then if n were 2, we could observe the values 0.00, 0.01,..., 0.98, 0.99. Therefore the discrete density function is given by (3).

$$P(\underline{r}_i = r_i) = 10^{-n} \quad \text{for} \quad r_i \in S$$
$$= 0 \quad \text{elsewhere} \tag{3}$$

where S is the collection of numbers $(j\ 10^{-n})$ with $j = 0, 1, \ldots, (10^n - 1)$. Hence if $n = 2$, then S consists of the numbers 0.00, 0.01, ..., 0.99. Also see Haitsma and Oosterhoff (1964, pp. 5-6), Halton (1970, p. 29), and Molenaar (1968, pp. 104-105). The cumulative distribution simply follows from the discrete density function in (3). This distribution is a stepwise approximation of the continuous distribution in (1). Figure 2 summarizes the various functions. When using a computer for the Monte Carlo calculations n is so high, or conversely 10^{-n} is so small, that we can neglect the difference between the continuous and the discrete distributions. It is convenient to work with the continuous form specified in (1).

(ii) With Fisz (1967, p. 55) we call the variables $\underline{r}_1, \underline{r}_2, \ldots, \underline{r}_i, \ldots, \underline{r}_n$ independent if their joint cumulative distribution, say G, can be written as the product of their individual distributions as in (4).

$$G(r_1, r_2, \ldots, r_n) = F_1(r_1)F_2(r_2) \ldots F_n(r_n) \tag{4}$$

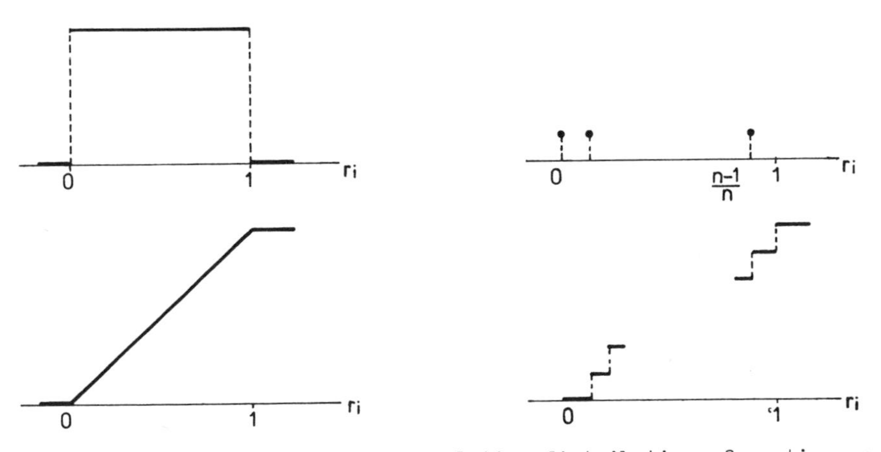

Fig. 2. Density function and cumulative distribution of continuous and discrete random numbers.

Since all n random numbers have the same distribution, specified in (1), we can write (4) as in (5).

$$G(r_1, r_2, \ldots, r_n) = F(r_1)F(r_2) \ldots F(r_n) \tag{5}$$

The independence of the random numbers implies that knowledge of $r_1, r_2, \ldots, r_{i-1}$ does not give any information about r_i. Physical processes that generate random numbers are described in Shreider (1964, pp. 220-238), Teichroew (1965, p. 40), and Tocher (1963, pp. 50-67). We mention as very simple physical devices a ten-sided die for throwing a decimal number between 0 and 9, and a coin for building up a binary number (which consists of 0's and 1's). As we shall see below these purely random numbers are unattractive compared with pseudorandom numbers.

Pseudorandom numbers are generated by applying a deterministic algebraic formula which results in numbers that for practical purposes are considered to behave as random numbers, i.e., to be also uniformly distributed and mutually independent. This independence implies that the "cycle length" is long enough. The cycle length is the number of pseudorandom numbers that is generated before we obtain the same sequences of numbers again. A much-used formula is the so-called multiplicative congruential or power residue method given in (6).

$$\begin{aligned} x_i &= a x_{i-1} \pmod{m} \qquad (i = 1, 2, \ldots) \\ x_0 &= b \end{aligned} \tag{6}$$

This equation specifies that the previous number x_{i-1} is multiplied by the constant a, the result is divided by m, and the remainder is taken as the new number x_i; the starting value or "seed" of x_i is b. The corresponding pseudorandom number r_i, which must be between 0 and 1, is obtained by dividing x_i by m. Pseudorandom numbers should be tested for uniformity and independence. Their cycle length can be determined from number theory. We shall return to pseudorandom numbers in Section I.6.4 to discuss some technical aspects and the overwhelming amount of literature on pseudorandom-number generation.

The advantage of pseudorandom numbers over purely random numbers is that the computer itself can generate pseudorandom numbers, using a formula like (6). In this way no storage of a large table of random numbers is needed, neither is it necessary for the computer to go through the slow process of reading in random numbers. In Chapter III we shall see that there is, moreover, the important statistical advantage of being able to reproduce the sequence of pseudorandom numbers. Such reproduction is possible since (6), for instance, shows that we generate the same sequence again if we start with the same initial value b, and the same a and m. Since, nowadays, computers always use pseudorandom numbers instead of purely random numbers, the term random number from now on stands for pseudorandom number in this study.

We have discussed the technical point of random numbers at some length since it is a basic concept in the definition of the Monte Carlo method. Moreover, later on we shall need several relations that were derived above. Next we shall examine three application areas of Monte Carlo.

(i) The first application area concerns the solution of deterministic problems applying the Monte Carlo method (in a wide sense), i.e., using random numbers. As Morgenthaler (1961, pp. 368-370) remarks the term Monte Carlo was introduced by von Neumann and Ulam in the late 1940's just for solving deterministic problems with the aid of random numbers which are stochastic variables. A deterministic problem can be solved by the Monte Carlo technique if the problem has the same formal expression as some stochastic process. As an example consider the integral

$$\int_{v}^{\infty} \frac{1}{x} \lambda e^{-\lambda x}\, dx \qquad (v, \lambda > 0) \tag{7}$$

The value of this integral can be estimated if we realize that $\lambda \exp(-\lambda x)$ for $x \geq 0$ is an exponential density function. Hence we can sample $\underline{x}$ from this density function and substitute the sampled value of $\underline{x}$ into $g(\underline{x})$ defined in (8).

$$\begin{aligned} g(\underline{x}) &= 0 \quad \text{if} \quad \underline{x} < v \\ &= \frac{1}{\underline{x}} \quad \text{if} \quad \underline{x} \geq v \end{aligned} \tag{8}$$

The expected value of $g(\underline{x})$ is equal to the integral in (7). This simple example is worked out in Appendix I.1. Another example is the calculation of the surface of an irregularly shaped figure in Morgenthaler (1961, p. 370). McCracken (1955) discusses the experimental calculation of the number π. More examples, e.g., multiple integrals, difference equations with side conditions, linear equations, and eigenvalues can be found in Mihram (1972, pp. 186-199), Newman and Odell (1971, pp. 53-57, 67-74), and Shreider (1964). Halton (1970) gives a very extensive survey (with 251 references!) of the use of the Monte Carlo method in solving problems in mathematics, physics, etc.

(ii) A second application area of the Monte Carlo method is formed by _distribution sampling_ which Morgenthaler (1961, p. 370) calls model sampling. Distribution sampling has been in use in _mathematical statistics_ for some decades. The purpose is to find the distribution or some of the parameters of the distribution of a stochastic variable. This stochastic variable, that we call the output variable, is a known function of one or more other stochastic input variables which have known distributions. To estimate the distribution of the output variable we draw a value for each of the input variables from their distributions and calculate the resulting value of the output variable. Such sampling is then repeated many times and this yields an estimate of the distribution. We have worked out a simple example of distribution sampling in Appendix I.2, where we estimate the probability that a stochastic variable $\underline{x}$ is smaller than some constant a, or

$$p = P(\underline{x} < a) \tag{9}$$

We assume that the output $\underline{x}$ is a function of the inputs $\underline{x}_1$ and $\underline{x}_2$ as specified in (10)

$$\underline{x} = \min(\underline{x}_1, \underline{x}_2) \tag{10}$$

and $\underline{x}_1$ and $\underline{x}_2$ have known distributions. Note that instead of the percentile p in (9) we may be interested in a related quantity, the "quantile," i.e., p in (9) is fixed (e.g., 5%) and we want to estimate the corresponding value of a; see Lewis (1972, pp. 9-10).

Other examples of distribution sampling are provided by the many Monte Carlo studies of the robustness of some statistic, a statistic being called more robust the less sensitive it is to violations of the underlying assumptions. Consider, for instance, the well-known t-statistic, defined in (11).

$$\underline{t} = \frac{\bar{\underline{x}} - \mu}{\underline{s}/\sqrt{N}} \tag{11}$$

We can use the Monte Carlo method to study the distribution of $\underline{t}$ if $\bar{\underline{x}}$ and $\underline{s}$ are based on, for instance, nonnormal observations $\underline{x}_i$. Andrews et al. (1972) made a Monte Carlo study of 70 different location estimators. Neave and Granger (1968) made a Monte Carlo study on the robustness of various tests for comparing two means. Donaldson (1966) investigated the robustness of the F-test. Heuts and Rens (1972) compared the power of two goodness-of-fit tests. In Chapter VI we shall present a Monte Carlo study on the robustness of a particular multiple ranking procedure. Many other studies, up to 1953, are mentioned by Teichroew (1965, p. 35). Lewis (1972) gives an excellent discussion of computer aspects of distribution sampling including more than 50 references, many of them very recent.

Other examples of distribution sampling are the many studies made in recent years on the "performance" of various regression analysis techniques for econometric models. In these studies a model is used to generate data; to these data various regression techniques are applied to estimate the parameters of the assumed model and the distributions of the resulting parameter estimators are compared with each other. Johnston (1963, pp. 275-295) describes several Monte Carlo studies on the performance of ordinary least squares, two-stage least squares, limited-information single equation and full-information maximum likelihood estimators applied to econometric models. Similar studies are made by Kmenta and Gilbert (1968), Mikhail (1972), Neeleman (1973), Sasser (1969), and Schink and Chiu (1966); also see the survey in Meier et al. (1969, pp. 138-141). Later on in this chapter we shall return briefly to distribution sampling in order to point out the similarities and dissimilarities of these sampling experiments and simulation.[2)]

(iii) A third application area of the Monte Carlo method is simulation. We shall define simulation in the next section. Here we remark only that many simulation studies also use random numbers and are therefore a form of Monte Carlo. We find it desirable to distinguish between simulation and other Monte Carlo applications. Therefore we shall speak of the Monte Carlo method in a narrow sense if we mean that part of the Monte Carlo (in a wide sense) that is no simulation. From now on we use the term Monte Carlo in this narrow sense. Monte Carlo stands for all solution techniques using random numbers, exclusive of simulation (see Fig. 5). This convention further implies that we do not follow some authors like Hammersley and Handscomb (1964) who let the term "Monte Carlo methods" comprise "variance reducing" techniques. These techniques will be discussed in Chapter III.

I.6. SIMULATION

I.6.1. Description of the Simulation Method

Naylor et al.(1967a, p. 2) and Tocher (1966, pp. 693-695) observe that there is no generally accepted definition of simulation. Following closely the definition in Naylor et al. (1967a, p. 3) we define simulation in a wide sense as "experimenting with a model over time." Let us have a closer look at this definition.

Simulation implies experimentation. However, instead of experimenting with the real world object we experiment by means of the model of that object. The behavior of the modeled object is followed over time. This description of simulation leads to two remarks. First, the object in a simulation study is a system. So Kiviat (1967, p. 5) states that it is common to find the terms "systems simulation" and "simulation" used interchangeably. Second, in Section I.3 we defined the term "model" in such a way that it is restricted to abstract models only. In the description of simulation we could drop this restriction and interpret the word model as both abstract and physical model. Examples of simulation with physical models would be provided by tank-tested scale models of ships, scale models of airplanes in wind tunnels, scale models for studying the influence of sea currents, etc. Some simulations with physical models might involve the participation of real people. Examples are link-trainers for pilots and automobile test tracks. Examples of simulation with abstract models, but also involving real people, would be business and military games. This kind of simulation is called operational gaming by Morgenthaler (1961, p. 37) and Naylor et al. (1967a, p. 3). For literature on games we refer to the bibliographies by Naylor (1969), Shubik (1960 and 1970), Shubik and Wolf (1972); also see Emshoff and Sisson (1971, pp. 245-249) and Meier et al. (1969, pp. 179-213). [The relations between gaming and mathematical game theory are discussed by Shubik (1972).] Simulation with physical or abstract models also involving real people, is called man-machine simulation; see, e.g., Geisler (1960) and Meier et al. (1969, pp. 287-289). In this study, however, we shall concentrate on simulation with abstract models only and without real people participating. Therefore, we defined simulation in a wide sense as experimenting with (abstract) models over time.

In the above definition simulation does not necessarily imply the use of random numbers. Deterministic simulation is quite common in economics. The economic model consists of regression equations that do include stochastic disturbance (or error) terms. When estimating the parameters of the regression equations these stochastic terms are accounted for. However, once the parameters are estimated the disturbances are suppressed and the timepath of the endogenous variables is calculated from the exogenous variables and lagged endogenous variables, without using random numbers. Examples are the simulation of the shoe-leather-hide industry by Cohen (1960) and the quarterly econometric model of the Dutch economy by Driehuis (1972). Howrey (1966) showed that this approach may yield false conclusions; also see Howrey (1972, p. 26), Kleijnen (1970, pp. 14-18), Naylor et al. (1967b, pp. 1314-1319) and Naylor et al. (1968, p. 188). A particular class of deterministic simulation is formed by Forrester's (1961) Industrial Dynamics approach, to which we shall briefly return in Section I.6.2. Other examples are Clough et al. (1965, p. 126) simulating the Canadian mining industry, and Molenaar (1968, p. 91-96) presenting a sequencing and queueing model. We recommend Emshoff and Sisson (1971, pp. 169-170) for a discussion of the choice between deterministic and stochastic simulation. Nevertheless, most simulation studies involve random numbers.

We define simulation in a narrow sense as experimenting with an (abstract) model over time, this experimentation involving the sampling of values of stochastic variables from their distributions. Therefore this simulation is called stochastic simulation. The technique of sampling stochastic variables using random numbers will be discussed in Section I.6.4. Because random numbers are used this type of simulation is sometimes denoted as Monte Carlo simulation. We shall discuss an example of stochastic simulation that will also be used in later chapters.

Suppose that we want to choose between two maintenance policies for a bus. The bus is scheduled to make N trips per day. Each day it starts in good condition. If the bus starts a particular trip in good condition it has a probability a of ending that trip in "deterioration phase A." In state A the bus can still run; repair is possible within the time needed for making one trip. If, however, the bus continues without repair after it has reached state A then it has a probability b of ending a trip in deterioration phase B. When it reaches state B all remaining trips of that day must be cancelled. Suppose we want to decide between strategy α, the bus is

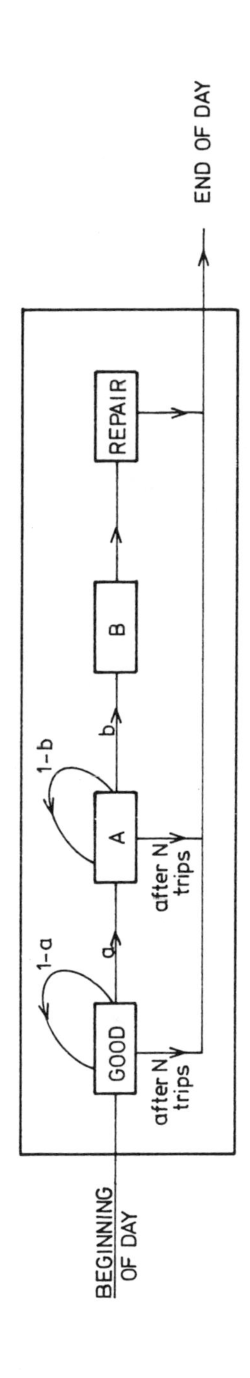

Fig. 3. Maintenance policy β.

repaired as soon as it gets in state A, and strategy β, the bus is not repaired until it reaches phase B. For strategy β the system is pictured in Fig. 3. This system can be simulated by using one random number $\underline{r}$ per trip. If the trip was started in good condition then we check if $\underline{r} < a$. If indeed $\underline{r} < a$ then that trip is ended in state A. Once the bus is in A we check if the relevant random number is smaller than b. If $\underline{r} < b$ then the bus is in B and all remaining trips must be cancelled since the bus is to be repaired. A new day can be simulated likewise. This example is worked out in Appendix I.3.

An example of a simulation that is often mentioned in the literature, is the simulation of a queuing system with one service station, a first-in-first-out (or FIFO) queuing discipline, and exponential interarrival and service times. If we sample the particular interarrival times between customers i and i + 1, say AT_i, and the service times of customer i, ST_i, shown in Fig. 4 then the waiting times (WT_i) and idle times of the service station (IT) shown in that figure result. We leave it to the reader to construct the appropriate flowchart for the simulation of this system.

There are many more examples of (stochastic) simulation. In the area of management science we mention simulation of complex queuing systems with priorities, several service stations and other complications that arise at traffic lights, airports, and steel plants (Naylor et al., 1967a; Schmidt and Taylor, 1970, pp. 325-403; Tocher, 1963). Mihram (1972, pp. 200-206) gives a classification of queuing systems based on arrival pattern, service mechanism, and queue

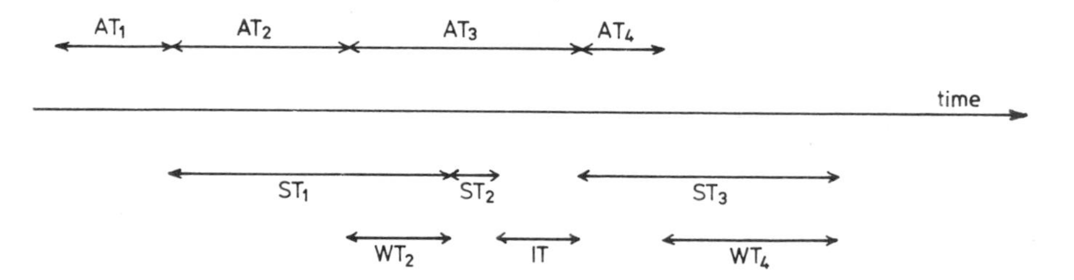

Fig. 4. Single-server FIFO queuing system.

discipline. A survey of simulation job shop scheduling is given by Moore and Wilson (1967) and Naylor (1971, p. 59). Simulation of computer systems is discussed by Bovet (1972), Dewan et al. (1972), Hutter (1970), Maisel and Gnugnoli (1972), Pomerantz (1970), and Rechtschaffen (1972, pp. 265-268); many more references are listed by Lucas (1971) and Maguire (1972). Inventory simulations are referenced by Naylor (1971, p. 53) and Schmidt and Taylor (1970, pp. 404-475). Queuing and inventory simulations are also surveyed by Smith (1968, pp. 67-69; and 1970, pp. 770-772). Mann (1970) discusses simulation (and other approaches) of systems reliability. Critical path calculations in networks are studied by Burt (1972), Burt et al. (1970), Klingel (1966), and Van Slyke (1963). An extensive discussion of marketing simulations, including a bibliography, is given by Elton and Rosenhead (1971). Financial (including capital budgeting) simulation models are surveyed by Naylor (1971, pp. 61-68); also see Benecke and Westgaard (1972), Carter (1971), and Salazar and Sen (1968). More applications can be found in textbooks and the proceedings of various conferences on simulation. Numerous abstracts of publications on simulation applications and methodology can be found in Computing Reviews and Operations Research/Management Science (published monthly) and International Abstracts in Operations Research (published bimonthly). The Dutch periodical Mededelingen Operationele Research (published monthly) also contains abstracts (most of them in English) on simulation. So it is not surprising that recent surveys, e.g., Shannon and Biles (1970, p. 744) and Turban (1972, p. 718), indicate that simulation is the operations-research technique most used in practice.

Another application area is formed by economics. Simulation models of the firm were constructed by Forrester using his industrial dynamics approach; by Bonini applying an interdisciplinary approach; by the Systems Development Corporation, etc. Cyert and March developed models for a duopoly and olipoly. A model for the lumber industry was constructed by Balderston and Hoggatt and, as we mentioned before, for the shoe, leather, and hide sequence by Cohen (1960).

Horn (1968) studied the cyclical movement in the supply of hogs in the Netherlands using both econometric regression models and simulation models. Models of the economy as a whole include the big Brookings-SSRC model for the USA, the model of Orcutt et al. (1961), and the model of Holland and Gillespie for India. These models and other economic models are discussed by Driehuis (1972), Dutton and Starbuck (1971a, pp. 465-588), Meier et al. (1969, pp. 118-138), Naylor et al. (1967a, pp. 192-233), Naylor (1971). We shall briefly return to simulation in economics in Section I.6.2.

Many large-scale simulations are done in the military field [see Koopman (1970), Morgenthaler (1961, pp. 393-395), Smith (1968, pp. 50-57), and many abstracts in Mededelingen Operationele Research]. For simulation of the heuristic way human beings solve problems, artificial intelligence, simulation of human thought, etc., we refer to Dutton and Starbuck (1971a) and Meier et al. (1969, pp. 147-178). Various application areas, especially in the behavioral sciences, are discussed by Emshoff and Sisson (1971, pp. 243-264). The above application areas and others are also examined in Guetzkow et al.(1972). Political simulations are collected in Coplin (1968) and Guetzkow (1971). Other application areas are demography, biology, metereology, in general any scientific discipline that uses quantitative relationships.

In Fig. 5 we have summarized some terminology. Notice that simulation in a narrow sense uses random numbers and therefore forms a subset of the Monte Carlo method in a wide sense. Some authors use the term Monte Carlo for what we call simulation (in a narrow sense) (Churchman et al., 1959, p. 174, 407; Jessop, 1956). In recent literature, however, the term simulation is more generally accepted (Mihram, 1972, pp. 207-208; Naylor et al. 1967a; Tocher, 1963).

I.6.2. Simulations and Systems

Since simulation involves experimentation over time the technique can be utilized for the study of the dynamic behavior of systems. Especially if the model contains stochastic variables, even a

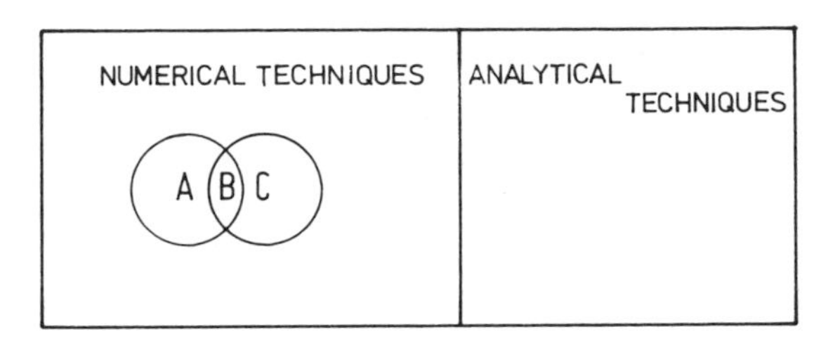

A + B = Monte Carlo in a wide sense

B + C = Simulation in a wide sense

A = Monte Carlo in a narrow sense = "Monte Carlo"

B = Simulation in a narrow sense

Fig. 5. Various solution techniques.

simple model may defy analytical study of the system behavior. We think that this holds even more for the study of transient behavior as opposed to steady-state behavior. For in the latter case limit theorems can be applied. It may be that we are not interested in the dynamic behavior of a system, but only in the result at the end of the timepath, e.g., the profit after 3 months. Even then simulation may be necessary if the system has variables which interact in a complicated way. Notice that we have come to realize that we should study the system as a whole rather than focus on the subsystems since the latter approach would lead to suboptimization. Unfortunately the determination of the overall optimum leads to very complicated models. Such a model may again not be solvable by analytical techniques, especially if some variables are stochastic.

Simulation is used for complicated systems. Often we know more about the behavior of the individual components of such a system than about the total system behavior. For instance, we know that if a customer leaves a service counter after being served, then the next customer who was standing in line will occupy that free counter; however, we do not know the relationship between "the" average service time and "the" average waiting time of a customer. The flowchart of the system shows which elements there are in the system, how the elements react, and how they influence each other. The mathematical relationships for each element are usually very simple; they are not necessarily linear equations but can be simple nonlinear equations or inequalities. Each element influences many other elements; it is these interactions that make the model complicated. If the model is based on our knowledge of the behavior of the components of the system instead of the total system then we speak of modular or building block designs or the reticulation method. The resulting models can be solved by simulation. So as Conway et al. (1959, pp. 93-94) state, simulation is possible if we know the (possibly probabilistic) behavior of the system elements.

In economics most models are not of the above building-block type. The traditional econometric models are sets of linear regression equations and are used to forecast the endogenous variables for the next period. More recently these models have been adapted, the adaptations consisting of introducing side conditions and nonlinearities expressing prior knowledge about the system. These models can be solved by simulation which, moreover, can produce complete time-paths (instead of a forecast for only the next period (Naylor, 1971). A modular model, however, would exist only if we used prior theoretical knowledge about the individual consumers and producers in order to explain macroeconomic relations. In this way we would not try to explain directly the total system behavior (as expressed by the relationships among the macroeconomic variables) but instead we would break down each macroeconomic variable into many microeconomic variables. For example, national consumption would be broken down into the consumption of individual 1, 2, etc. Such an approach was actually applied by Orcutt et al. (1961) in what was meant to be an economic study, but what turned out to be a demographic study. A discussion of the differences between traditional econometric models and simulation models can also be found in Cohen (1960), Kleijnen (1970), and Reith (1969). Howrey and Kelejian (1969) wrote an excellent paper in which they compare analytical solutions for economic models with simulation. They are very skeptical about the utility of simulation in economics. Their conclusion, however, is not based on what in our opinion is the most important application of simulation, i.e., stochastic simulation of nonlinear models. Recently economists are showing increasing interest in Forrester's industrial dynamics approach. As we observed in Section I.6.1 this approach is (for the most part) deterministic simulation. It allows the modeling of subsystems that can be connected together. Characteristic for it are the difference equations that form the model (i.e., the system is evaluated at fixed intervals of time) and the high degree of aggregation. For a further discussion of industrial dynamics and its differences with simulation as used in operations research, we refer to Meier et al. (1969, pp. 80-117, 277-282) and also to Niedereichholz (1971) and Smith (1968, pp. 27-32).

Let us return to distribution sampling discussed in Section 1.5. Teichroew (1965, p. 43) suggests characteristics that distinguish distribution sampling from simulation:

(i) The output variable in distribution sampling is a relatively simple function of the input variables and the parameters. In simulation this is a very complicated function which cannot be formulated explicitly otherwise than by the whole computer program!

(ii) The function in distribution sampling is static, whereas in simulation the relations are dynamic. We observe that studies on the performance of regression techniques were mentioned as examples of distribution sampling. These studies aim at finding the distribution of the estimators of the parameters of an econometric model. Such studies have the same purpose as other examples of distribution sampling, and in common with simulation studies, they may use complicated and dynamic relationships.

We have seen that most simulation models are based on our knowledge of the behavior of the system elements. The individual behavior can often be represented by simple operations. (It is the numerous interactions that make the total system complicated.) The simulation experiment is carried out by executing these simple operations many times. For example, the arrival and servicing of a customer is executed a great many times. Since simulation is based on the repetition of many operations, it is well-fitted for a digital computer. Programming problems in digital simulation are indicated by Conway et al. (1959, pp. 95-103) and Morgenthaler (1961, pp. 405-411), and some useful programming hints are given by Hauser, et al. (1966, pp. 84-85). For a discussion of digital and analog simulation of continuous systems we refer to Bauknecht (1971, pp. 157-160), Emshoff and Sisson (1971, p. 9), Mihram (1972, pp. 226-228), Morgenthaler (1961, pp. 376-378, 384-385), Nisenfeld (1971), Smith (1968, pp. 15-26) and van Dixhoorn and Lyesen (1968, pp. 4-6, 35-39). Many programming problems have been eliminated by the current development of special simulation languages. A survey of these languages and more references can be found in Bauknecht (1971), Emshoff and Sisson (1971, pp. 115-158), Kay (1972), Maguire (1972), Maisel and Gnugnoli (1972), Meier et al. (1969), Mihram (1972, pp. 238-244, 493-495), Naylor et al. (1967a), Naylor (1971), and Rytz (1971).

I.6.3. Problems in Simulation

Simulation does have its drawbacks. Even on a modern computer it may take much time to execute one "run," i.e., one movement of the system from the beginning to the end of the simulated time period. Moreover, since it is an experiment it gives output only for the particular values of the independent variables and the system parameters used in that experiment. (System parameters are those characteristics of the system that remain constant during a run.) So simulation does not provide functional relationships between the output and the independent variables and parameters. Therefore a "sensitivity analysis" for the parameters requires new runs with one or more parameters set a new values. "Sensitivity" experiments are performed to determine if a change in a particular parameter strongly influences the output. (If this is the case then it is important to obtain more information about the true value of that parameter; so we may then decide to spend more money for acquiring additional information about that particular parameter.) The above argument implies that an optimal solution for the system can only be found experimentally, i.e., we have to try out various parameter sets in order to approximate the optimum set. In simulation in a narrow sense stochastic fluctuations further complicate the determination of the optimum solution. Fortunately there are techniques for systematically searching for the optimum, e.g., "response surface methodology," which will be discussed in later chapters. (There are, however, problems that are not suited for these techniques. Some situations are characterized by combinatorial problems; cf. scheduling models. We can try to find a solution for this type of problems using heuristics; see Meier et al. (1969, pp. 147-178).)

Let us have a closer look at stochastic simulation. In this type of simulation we sample values of the stochastic variables in the model. Therefore, it is actually a statistical sampling experiment with the model of a system. Hence stochastic simulation has additional disadvantages because of its sampling character. This sampling involves all the problems of statistical design and analysis. (The design and analysis are often complicated by the special nature of the simulation experiment leading e.g. to autocorrelated data.) These statistical aspects of simulation will be surveyed in Chapter II and discussed in more detail in the other chapters.

For a further discussion of the disadvantages (and advantages) of simulation in a narrow and wide sense, we refer to the literature, e.g., Dutton and Starbuck (1971a, p. 705), Hillier and Lieberman (1968, pp. 470-471) and Naylor et al. (1967a, pp. 4-9). Even while simulation in a narrow or wide sense has certain disadvantages it remains a very important solution technique. Morgenthaler (1961, pp. 366-367, 372-375) lists eighteen reasons for using simulation! Here we point out only that researchers are constructing increasingly complicated models. Such models are needed, since, as we observed before, the total system with all its interactions must be considered in order to avoid suboptimization. These models can often be solved only by simulation; more powerful modern computers and special simulation languages make it possible to make even more use of this technique. Hence it is not surprising that Kiviat (1967, p. 5) states that "most of today's complex engineering and management studies include simulation experiments." The many examples of simulation referred to above, further illustrate the wide application of this technique.

I.6.4. Generation of Random and Stochastic Variables

In Section I.5 we briefly discussed pseudorandom numbers, and mentioned as an example the multiplicative congruential generator. Other types of generators (e.g., additive, mixed multiplicative, midsquare) can be found in the literature; see Mihram (1972, pp. 44-57), Naylor et al. (1967a, pp. 45-47) and Tocher (1963, pp. 72-84) for a nonsophisticated survey. The choice of the parameter values of these generators has several aspects. Number theory can be used to select such values that a long cycle length results. Mathematical statistics provide some guidelines for selection of parameters such that first-order serial correlation of the random numbers (i.e., correlation between $\underline{r}_i$ and $\underline{r}_{i+1}$) is minimized. These guidelines, however, are necessary but not sufficient conditions; moreover, they do not protect against serial correlations of higher order. Guidelines for parameter selection can be found in J.M. Chambers (1970) and R. P. Chambers (1967), Jansson (1966, pp. 39-68), Knuth (1969, pp. 9-33), Newman and Odell (1971, pp. 7-17); in textbooks on simulation, e.g., Mihram (1972), Naylor et al. (1967a), Naylor (1971), Schmidt and Taylor (1970), and Tocher (1963). Note that Marsaglia (1972) rejects all previous recommendations and gives a new table with recommended values for the multiplicative constant a in (6) above, when the modulus m is 2^{32}, 2^{35}, or 2^{36}. (It is convenient to take m equal to a power of 2, possibly ± 1, for binary computers, and equal to a power of 10 for decimal computers.) The recommended parameters do not guarantee truly random numbers. Therefore it is necessary to test the output of the generator for uniformity and independence. The above references also discuss a number of such statistical tests [also see Newman and Odell (1971, pp. 75-81)]. Unfortunately these tests check dependence of low order only, and as it is pointed out by Lewis (1972) and Marsaglia (1972) the dependence in higher dimensions may create problems. This high-order dependence may be mitigated by shuffling the random numbers before use; see Andrews et al. (1972, p. 63, 306), Lewis (1972), Marsaglia (1972), Marsaglia et al. (1972), Marsaglia and Bray (1968). Coveyou and MacPherson (1967) applied Fourier analysis to obtain a priori predictions of the statistical behavior of various generators; also see Knuth (1969, pp. 82-97) and Mihram (1972, pp. 473-477). This Fourier or spectral analysis itself will be discussed in some more detail in Section II.10. Here we state only that these authors do not apply statistical tests to the experimental output of the generators but give a mathematical derivation of the statistical performance of the generators. Coveyou and MacPherson (1967, p. 119) conclude that there "is at present no method of generation of pseudouniform sequences better than the simple multiplicative congruence method with a carefully chosen multiplier." Lewis et al (1969) give such a generator which indeed passed all their tests and is very efficient for the IBM System 360. A very fast version of their generator was recently derived and is reproduced in Appendix I.4. Marsaglia (1972), however, emphatically rejects Coveyou and MacPherson's analysis and shows that Lewis and associates used a multiplier that does not pass Marsaglia's tests; as we mentioned above Marsaglia gives a table of multipliers he recommends.

Our conclusion is that after so many years of research on pseudorandom number generators there still is no foolproof generator. At present the best one seems to be the multiplicative generator. However, a careful choice of the multiplier is necessary since otherwise unsatisfactory results are produced; as Lewis (1972, p. 5-8) and Marsaglia (1972) point out, many popular generators actually are unacceptable. Lewis (1972) recommends the Lewis-Goodman-Miller generator; Marsaglia gives his table of recommended multipliers; but even their output should best be shuffled before use. The best references for further study are Lewis (1972), Marsaglia (1972), and the bibliographies shown in Halton (1970, p. 32-47) and Nance and Overstreet

(1972). The latter bibliography gives 491 references on random number generators, their tests and their use in sampling from other distributions. Sampling stochastic variables from an arbitrary distribution is the second topic of this section.

Sampling values of a stochastic variable using random numbers is basic for stochastic simulation and later on we shall use some relations derived below. For a very simple exposé on the nature of this sampling process we refer to Flagle (1960, pp. 428-435). We shall discuss a widely applicable procedure.

Denote the cumulative distribution of the stochastic variable $\underline{x}$ by $F(x)$. So

$$P(\underline{x} < x) = F(x) \tag{12}$$

If $F(x)$ is a continuous, increasing function[3] then the inverse function exists and may be denoted by $F^{-1}(x)$. (Compare, if y is a continuous increasing function of z, say $y = g(z)$, then we can inversely express z as a function of y by $z = g^{-1}(y)$, i.e., $y = g(z)$ is solved for z.) Now we show that a value of $\underline{x}$ can be sampled from its distribution $F(x)$ by using the inverse F^{-1} and a random number $\underline{r}$, as specified in (13).

$$\underline{x} = F^{-1}(\underline{r}) \tag{13}$$

For, from (13) it follows that

$$P(\underline{x} < x) = P[F^{-1}(\underline{r}) < x] = P[\underline{r} < F(x)] = F(x) \tag{14}$$

the last equality resulting from (1). Comparing (14) with (12) shows that (13) gives the desired result. We shall illustrate the inversion technique with an example (that will again be used later on).

Consider the exponential distribution which is often met in simulation studies. The variable $\underline{x}$ is <u>exponentially</u> distributed if its density function, say, $f(x)$ is given by (15).

$$\begin{aligned} f(x) &= \lambda e^{-\lambda x} \quad \text{for } x \geq 0 \\ &= 0 \quad \text{for } x < 0 \end{aligned} \tag{15}$$

where $\lambda > 0$. Hence the cumulative distribution is given by (16).

$$F(x) = \int_0^x \lambda e^{-\lambda t}\, dt = 1 - e^{-\lambda x} \quad (x \geq 0) \tag{16}$$

Applying (13) means that we put $F(x)$ equal to the random number $\underline{r}$ and solve for $\underline{x}$, i.e.,

$$1 - e^{-\lambda \underline{x}} = \underline{r} \tag{17}$$

or

$$\underline{x} = -\frac{1}{\lambda}\, \ell n(1 - \underline{r}) = -\ell n(1 - \underline{r})/\lambda \tag{18}$$

So (18) can be used to generate exponential variates. Usually a slightly different formula is applied. For its derivation we have to show that the distribution of $(1 - \underline{r})$ is the same as that of $\underline{r}$ itself. This can be proved using the general formula for the density function of a transformed stochastic variable given, for instance, in Fisz (1967, p. 39). It can also be shown simply as follows. Denote $(1 - \underline{r})$ by $\underline{z}$. Then (19) holds.

$$\begin{aligned} P(\underline{z} < z) &= P(1 - \underline{r} < z) = P(\underline{r} > 1 - z) \\ &= 1 - P(\underline{r} < 1 - z) = 1 - (1 - z) = z \end{aligned} \tag{19}$$

where $0 \leq z \leq 1$. Comparing (19) with (1) shows that $\underline{z}$ $(= 1 - \underline{r})$ has the same uniform distribution as $\underline{r}$. So in (18) we can replace $(1 - \underline{r})$ by $\underline{r}$. This results in (20).

$$\underline{x} = -\ell n(\underline{r})/\lambda \tag{20}$$

We prefer (20) over (18) since (18) needs more computer time as it contains one more subtraction operation. In Chapter III the fact that $\underline{r}$ and $(1 - \underline{r})$ have the same distribution will be used again to derive the "antithetic" variance reducing technique.

The inversion technique can be applied to other distributions. For some distributions it is necessary or more efficient to use <u>other techniques</u>. For example, in Appendix I.2 we generate a normal variate using the central limit theorem. A general discussion of the inversion and other techniques can be found in Chambers (1970, pp. 6-10), Halton (1970, p. 29-32), Knuth (1969, pp. 100-121), Kohlas (1971, pp. 20-25), Mihram (1972, pp. 37-39, 94-146), Naylor et al. (1967a, pp. 68-122), Newman and Odell (1971, pp. 18-52), Schmidt and Taylor (1970, pp. 258-324), and Tocher (1963, pp. 6-42). These publications discuss sampling from numerous types of distributions, e.g., normal, χ^2, F, t, β, γ, Poisson, lognormal, multivariate normal, Wishart, time series. Some books, e.g., Mihram (1972) and Naylor et al. (1967a) also show the application areas of these types of distribution. Special distribution types are discussed in Ahrens and Dieter (1972) and Marsaglia et al. (1972) (normal and exponential distributions), Andrews et al. (1972, pp. 56-57) (sample $\underline{x}$ as $\underline{z}/\underline{y}$ where $\underline{z}$ is normal and appropriate choice of $\underline{y}$ yields various distributions), van Doeland and van Daal (1972) (exponential with changing parameter; also see van Daal and van Doeland (1972), Ten Broeke (1972), van Daal (1972)). In Appendix I.5 we discuss sampling two correlated variables as it might be desirable in capital investment simulation; compare Benecke and Westgaard (1972). In Section VI.3 we shall generate linear combinations of exponential variates to create distributions with the desired characteristics.

I.7. LITERATURE

Of the many references listed at the end of this chapter we would recommend Naylor et al. (1967a) as an excellent textbook giving an introduction to all aspects of simulation. Emshoff and Sisson (1971), Meier et al. (1969), and Tocher (1963) also published good books on simulation. Other excellent introductions, not as extensive as these books, are given by Flagle (1960), Hillier and Lieberman (1968), Morgenthaler (1961), Nelson (1966), and Sisson (1969). Among the other references there are some very good studies which, however, concentrate on more specialistic topics and are not mentioned here.

For a study of simulation in depth we refer to bibliographies on simulation and gaming drafted by Dutton and Starbuck (1971a, pp.

9-102, 693-699, 701-708), rearranged in Dutton and Starbuck (1971b) and Naylor (1969). These bibliographies are most recent, very extensive (over 2000 references in Dutton and Starbuck) and give some more bibliographies. For the sake of completeness we mention bibliographies they do not list, namely, the Netherlands A.D.P. Research Center (1967) and Shubik (1970); recently Mihram (1971, pp. 43-44) presented a bibliography of simulation bibliographies.

GLOSSARY

This glossary gives only a rough description of those few concepts that are needed in the rest of the book. A recent complete glossary can be found in Mihram (1971).

Distribution sampling	Monte Carlo application in mathematical statistics estimating the distribution or some parameters of the distribution of a stochastic variable
Model	Mathematical symbols or flowcharts representing real-world system
Monte Carlo in a narrow sense	Monte Carlo in a wide sense excluding simulation
Monte Carlo in a wide sense	Any technique for the solution of a model using <u>random</u> (or pseudorandom) <u>numbers</u>
(Pseudo) random numbers	Stochastic variables uniformly distributed on the interval [0,1] and independent
Simulation in a narrow sense	Experimenting with a model over time involving sampling of stochastic variables
Simulation in a wide sense	Experimenting with a model over time
Steady state	The system is in its steady state if its probability distribution does not vary over time
System	A set of elements having certain characteristics or attributes that assume numerical or logical values; there are relationships among the elements within the system and with the environment outside the system.

APPENDIX I.1. MONTE CARLO ESTIMATION OF THE VALUE OF AN INTEGRAL

In this appendix we shall give an example of the use of the Monte Carlo method for solving a deterministic problem. Suppose that we want to determine the value of the integral in (1.1), this value being denoted by $\xi(\lambda,v)$ or briefly by ξ.

$$\xi(\lambda,v) = \int_v^\infty \frac{1}{x}\lambda e^{-\lambda x}\,dx \quad (\lambda,\ v > 0) \tag{1.1}$$

The integral in (1.1) looks quite simple, but nevertheless it cannot be solved by straightforward application of partial integration or serial expansion. It can be shown that (1.2) holds.[4)]

$$\xi(\lambda,v) = \lambda\left[-\ c\ -\ \ell n(\lambda v) + \sum_{i=1}^{\infty}\left\{(-1)^{i+1}\frac{(\lambda v)^i}{i!i}\right\}\right] \tag{1.2}$$

where c is the Euler constant. We imagine that many nonmathematicians will have problems in deriving (1.2). However, Monte Carlo estimation of (1.1) is very simple. This estimation can be performed as follows.

(i) Sample a value of $\underline{x}$ from its exponential density function

$$\begin{aligned} f(x) &= \lambda e^{-\lambda x} && \text{for } x \geq 0 \\ &= 0 && \text{for } x < 0 \end{aligned} \tag{1.3}$$

This sampling can be done using (20).

(ii) Substitute the sampled value of $\underline{x}$ into $g(\underline{x})$ defined in (1.4)

$$\begin{aligned} g(\underline{x}) &= 0 && \text{if } \underline{x} < v \\ &= \frac{1}{\underline{x}} && \text{if } \underline{x} \geq v \end{aligned} \tag{1.4}$$

The expected value of $g(\underline{x})$ is given by (1.5).

$$\begin{aligned} E[g(\underline{x})] &= \int_{-\infty}^{\infty} g(x)\ f(x)\ dx \\ &= 0 + \int_v^\infty \frac{1}{x}\lambda e^{-\lambda x}\,dx = \xi(\lambda,v) \end{aligned} \tag{1.5}$$

(iii) Repeat these two steps a number of times using different random numbers. If $\underline{x}_i$ denotes the observation on $\underline{x}$ sampled in replication i $(i = 1, \ldots, n)$, then ξ can be estimated by $\hat{\underline{\xi}}$ defined in (1.6).

$$\hat{\underline{\xi}} = \frac{1}{n} \sum_{i=1}^{n} g(\underline{x}_i) \tag{1.6}$$

Because of (1.5) $\hat{\underline{\xi}}$ is an unbiased estimator of ξ. As n increases, the variance of this estimator decreases, therefore the probability of correct estimation of ξ increases. In Chapter III we shall return to this example to show how more refined sampling can be performed using a variance reduction technique called importance sampling.[5)]

We remark that the integral in (2.1) may be needed in certain inventory systems. For in Kriens and de Leve (1966, p. 71) it can be found that for the optimization of a particular inventory system (1.7) has to be solved for v.

$$\int_0^v f(x)\, dx + v \int_v^\infty \frac{1}{x} f(x)\, dx = \frac{C_2}{C_1 + C_2} \tag{1.7}$$

where v = initial inventory after receiving the ordered quantity
$\underline{x}$ = stochastic demand per unit of time
$f(x)$ = probability density function of demand per time unit
C_1 = holding cost per unit of goods for a time unit
C_2 = shortage cost per unit of goods for a time unit

If $f(x)$ is an exponential density function then (1.7) involves the determination of $\xi(\lambda,v)$ defined in (1.1) above.[6)] Notice that in this problem formulation the integral in (1.1) has a probabilistic interpretation at once. However, more generally speaking, we can use the above Monte Carlo procedure for an integral with an integrand, say, $h(x)$ that can be written as $h_1(x)\, h_2(x)$ where $h_2(x)$ is a probability density function from which we sample $\underline{x}$ and substitute into $h_1(\underline{x})$.

APPENDIX I.2. AN EXAMPLE OF DISTRIBUTION SAMPLING

In this appendix we shall use the Monte Carlo method to estimate p where

$$p = P(\underline{x} < a) \tag{2.1}$$

$$\underline{x} = \min(\underline{x}_1, \underline{x}_2) \tag{2.2}$$

while $\underline{x}_1$ and $\underline{x}_2$ are assumed to have independent normal distributions with mean 100 and variance 400 for $\underline{x}_1$ and mean 90 and variance 100 for $\underline{x}_2$, i.e.,

$$\underline{x}_1 : N(100,400) \tag{2.3}$$

$$\underline{x}_2 : N(\ 90,100) \tag{2.4}$$

The problem of the determination of p arises if we have a product consisting of two parts. The life of part 1, say $\underline{x}_1$, is determined by (2.3) the life of part 2 by (2.4). The product breaks down as soon as one of the two parts fails, i.e., the life of the product, $\underline{x}$, is determined by (2.2). We want to know the probability that the life of the product is smaller than some given value, say, a. So we want to know p defined in (2.1). This physical interpretation can also be found in Churchman et al. (1959, p. 174).

For the Monte Carlo estimation of p it is convenient to introduce the variable $\underline{y}$ defined in (2.5).

$$\begin{aligned} \underline{y} &= 1 \quad \text{if} \quad \underline{x} < a \\ &= 0 \quad \text{if} \quad \underline{x} \geq a \end{aligned} \tag{2.5}$$

The expected value of $\underline{y}$ is equal to p defined in (2.1) for

$$\begin{aligned} E(\underline{y}) &= 1 \cdot P(\underline{x} < a) + 0 \cdot P(\underline{x} \geq a) \\ &= P(\underline{x} < a) = p \end{aligned} \tag{2.6}$$

We do not generate <u>normal</u> variates applying the inversion technique of (13). Instead we make use of the <u>central limit theorem</u> which implies that the sum of a "large" number of independent stochastic variables with the same distribution and a finite mean and standard deviation is approximately normally distributed; see Fisz (1967, p. 197). So we add twelve random numbers assuming that twelve is high enough to yield an (approximately) normally distributed variable. Notice that the random number $\underline{r}$ has mean 0.5 and variance 1/12. Hence $\underline{z}$ defined in (2.7) has mean 0 and variance 1 and if the central limit theorem holds, this $\underline{z}$ is normally distributed.

$$\underline{z} = \sum_{j=1}^{12} \underline{r}_j - 6 \tag{2.7}$$

If $\underline{z}$ has density $N(0,1)$, then we know that $\sigma z + \mu$ has density $N(\mu, \sigma^2)$. So to generate $\underline{x}_1$ and $\underline{x}_2$ defined in (2.3) and (2.4), we use the relations (2.8) and (2.9), respectively.

$$\underline{x}_1 = 20\underline{z} + 100 \tag{2.8}$$

and

$$\underline{x}_2 = 10\underline{z} + 90 \tag{2.9}$$

Observe that in order to keep $\underline{x}_1$ and $\underline{x}_2$ independent we have to use different values of $\underline{z}$ in (2.8) and (2.9). A new independent value of $\underline{z}$ requires the generation of twelve new random numbers in (2.7). Therefore we introduced a switch, namely, the variable I, in Fig. 2.1. The sampling procedure is further specified in Fig. 2.1. Because of (2.6) we estimate p by (2.10)

$$\hat{\underline{p}} = \bar{\underline{y}}. = \sum_{j=1}^{N} \underline{y}_j / N \tag{2.10}$$

where $\Sigma \underline{y}_j$ is denoted by SUMY in the flowchart.[7] We assume that the number of values of $\underline{y}$ we want to sample, is NMAX. In Chapter V we shall derive how we can determine NMAX in such a way that we are 95% confident that our estimate is less than 10% wrong.

We remark that for the Monte Carlo estimation of p it is crucial that the distribution of the input variables $\underline{x}_1$ and $\underline{x}_2$ is given. However, it is not necessary at all that this distribution has the form assumed in our problem formulation. Two independent normal distributions are specified only to simplify the analytical determination of p which was used to check the Monte Carlo estimate of p.

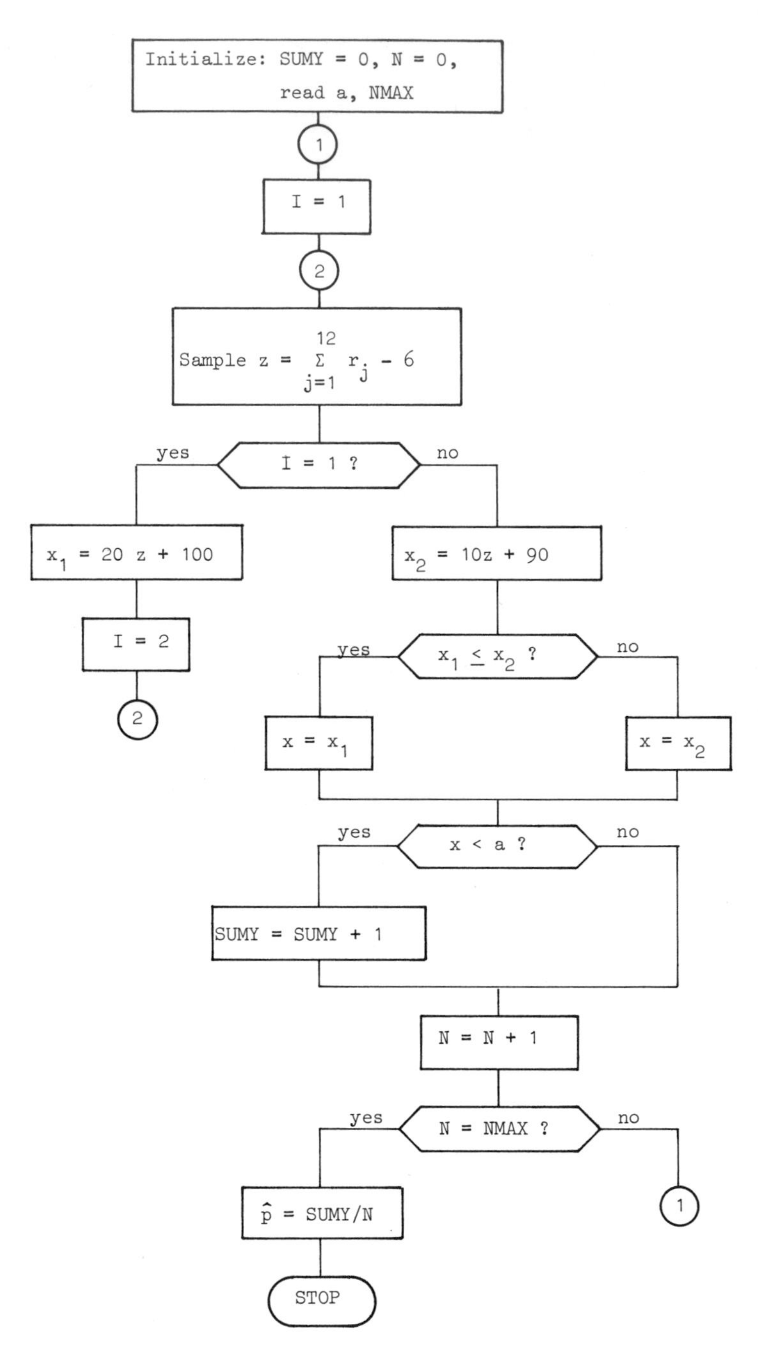

Fig. 2.1. Monte Carlo estimation of the fraction p.

APPENDIX I.3. SIMULATION OF MAINTENANCE STRATEGIES

The example we shall discuss in this appendix is the problem of the choice between two maintenance policies. This problem was formulated above; see Fig. 3 and the relevant text. For the convenience of the reader we shall repeat the problem once more. Suppose a bus is scheduled to make N trips per day. The bus is subject to wear and tear, so _repair_ is necessary. There are two phases in this deterioration process. When the bus is in _state A_ it can still run, but it will eventually need repair. If the bus continues without repair after state A then the bus might get in _state B_. In state B the bus is in such a bad condition that it has to discontinue service immediately. The "_transition probabilities_" are supposed to be as follows. If the bus starts a trip in good condition, then it has a probability (a) of ending that trip in state A and a probability (1 - a) of remaining in good condition. If the bus starts a trip in state A, then it has a probability (b) of ending the trip in state B and a probability (1 - b) of remaining in state A. The _duration_ of the repair varies. For, if the bus is repaired in state A, then repair is simple and takes the time needed for making a single trip. So one trip is cancelled. However, if the bus reaches state B, then all remaining trips of that day must be cancelled and the bus is not available until the next day. Suppose we want to choose between the following two maintenance strategies.

Strategy α: As soon as the bus is in state A, it is repaired.

Strategy β: The bus keeps running until it is in state B.

The strategy with the highest mean number of trips per day is supposed to be preferred. Finally, we suppose that the bus starts in good condition each day. So, if under strategy α or β the bus ends the Nth trip in state A or B, then it is repaired overnight.

Above we formulated the case as a bus-maintenance problem. It will be clear that we could also have presented it as a machine-maintenance problem. This model was originally given by Morse (1962, pp. 98-100). We, however, used the formulation of Kriens (1964, p. 31).

In both publications the case is solved applying Markov theory. These two authors derived that the expected number of trips per day under strategy α and β, respectively, is

$$\mu_\alpha = \frac{N}{1 + a} + a\frac{1 - (-a)^N}{(1 + a)^2} \tag{3.1}$$

$$\mu_\beta = \frac{a^2[1 - (1 - b)^N] - b^2[1 - (1 - a)^N]}{ab(a - b)} \tag{3.2}$$

We shall now show how this problem can be solved in a simple way by simulation.

In the simulation program we sample the state in which the bus is at the end of a trip by generating a random number $\underline{r}$ and using Table 3.1. For example, suppose the bus starts a trip in state A. (Notice that starting in state A is possible only under strategy β.) If the generated value of $\underline{r}$ satisfies $\underline{r} \geq b$, then we have the bus end that trip in state A. For

$$P(\underline{r} \geq b) = 1 - P(\underline{r} < b) = 1 - b \tag{3.3}$$

So (3.3) demonstrates that this sampling procedure satisfies the transition probabilities in the given model.

In Fig. 3.1 the flowchart for the simulation model is given. We remark that it would have been possible to simulate first the operation of the bus during, say, M days under strategy α and then during M days under strategy β. In Fig. 3.1, however, we first have the bus make one trip under strategy α and then under strategy β; next the following trip is made, again first under α and then under β,

TABLE 3.1. Sampling the State of the Bus at the End of a Trip

Bus starts trip in state	Bus ends trip in state		
	Good	A	B
Good	$a \leq \underline{r} < 1$	$0 \leq \underline{r} < a$	
A	-	$b \leq \underline{r} < 1$	$0 \leq \underline{r} < b$

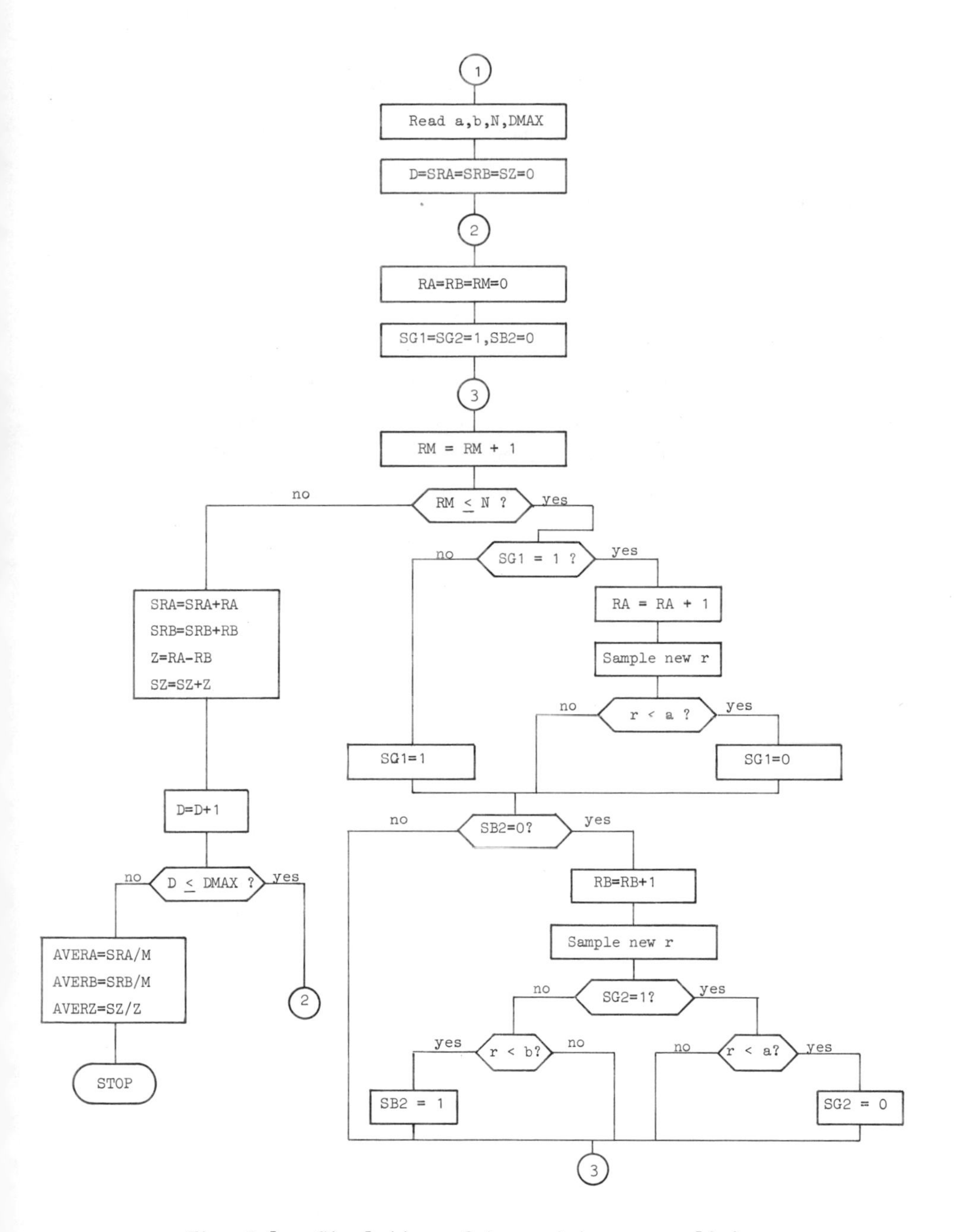

Fig. 3.1. Simulation of two maintenance policies.

etc. This procedure was chosen in order to apply the "common random numbers" technique in a simple way. This technique will be presented in Chapter III were we shall return to this maintenance problem.

In the flowchart of Fig. 3.1 the following mnemonic symbols are used that are not yet defined.

SG1: the (logical) variable "state good under 1st strategy" has the value 1 if the bus is in good condition under strategy α; it has the value 0 if the bus is in state A under strategy α.

SG2: the (logical) variable "state good under 2nd strategy" has the value 1 if the bus is in good condition under strategy β; the value 0 if it is in state A.

SB2: the variable "state B under 2nd strategy" has the value 1 if the bus is in state B under strategy β; otherwise it is 0.

DMAX: number of days we want to simulate.

D: number of days simulated so far.

RM: number of simulated trips per day, inclusive of cancelled trips.

RA,RB: number of trips that the bus has actually made during a day under strategy α and β, respectively.

SRA,SRB: number of trips that the bus has made during D days under strategy α and β, respectively.

Z: difference between trips actually made under strategy α and β on a certain day, i.e., Z = RA - RB.

SZ: Sum of Z's over D days.

AVERA,AVERB: averaged number of trips per day under strategy α and β respectively, after M days.

AVERZ: AVERA - AVERB.

Notice that in Fig. 3.1 the part of the flowchart between the connectors 2 and 3 is used only once on each simulated day; at the beginning of a day no trips have been made (RA = RB = 0) or simulated (RM = 0) and the bus starts in good condition under both strategies (SG1 = 1, SG2 = 1, SB2 = 0). After connector 3 RM is compared with N to check if the maximum number of trips per day, N, has been simulated already. If under strategy α the bus starts a trip in good condition (SG1 = 1) then the trip is made (RA = RA + 1) and the state at the end of that trip is sampled, i.e., if $\underline{r} < a$ then the bus is no longer in good condition (SG1 = 0); if $\underline{r} \geq a$ then no change occurs. If the bus is not in good condition at the beginning of a trip (SG1 = 0), then that trip is not made (no RA = RA + 1), but the bus is repaired (SG1 = 1). We do not need to sample a random number, for the bus will definitely be in good condition after the repair! Next we determine if a trip is made under strategy β and, using Table 3.1, in which state a trip is ended. If one day has been simulated (RM = N), then we update SRA, SRB and SZ; we simulate a next day (D = D + 1) if the required number of days (DMAX) has not yet been simulated. After DMAX simulated days we calculate the average of trips per day under strategy α and β and their difference.

We remark that with the simulation method changes in the model can often be realized by simple changes in the flowchart. Suppose, for instance, we drop the assumption that under strategy β the bus is always in good condition at the beginning of the day. So we actually consider a strategy γ where a bus that ends a day in state A is not repaired overnight. We leave it to the reader to make the few necessary adjustments in the flowchart of Fig. 3.1.

APPENDIX I.4. LEWIS-LEARMONTH RANDOM NUMBER GENERATOR[8)]

```
********************************************************************** RAND0001
*                                                                      RAND0002
*   LEWIS-LEARMONTH UNIFORM RANDOM NUMBER GENERATOR                    RAND0003
*   NAVAL POSTGRADUATE SCHOOL MONTEREY CALIFORNIA                      RAND0004
*                                                                      RAND0005
********************************************************************** RAND0010
*                                                                      RAND0020
*   SUBROUTINE: RANDOM  (ADDITIONAL ENTRY POINT: OVFLOW)               RAND0030
*                                                                      RAND0040
*   USAGE: RANDOM IS A PSEUDO-RANDOM GENERATOR FOR UNIFORMLY DISTRIBUTED RAND0050
*          VARIATES.  RANDOM RETURNS BOTH INTEGER*4 VARIATES AND REAL*4 RAND0060
*          VARIATES.  FOR INTEGER VARIATES, RANDOM RETURNS ONE VARIATE RAND0070
*          ON THE CLOSED INTERVAL 1 TO 2147483647 PER CALL. FOR REAL*4 RAND0080
*          VARIATES,RANDOM RETURNS AS MANY VARIATES AS DESIRED (SEE    RAND0090
*          CALLING SEQUENCE ) ON THE OPEN INTERVAL 0.0 TO 1.0.  TO     RAND0100
*          AVOID THE TIME CONSUMING PROCESS OF DIVISION BY THE MODULUS, RAND0110
*          RANDOM REQUIRES A SPECIAL CALL TO ITS ENTRY POINT OVFLOW TO RAND0120
*          INTERCEPT FIXED-POINT OVERFLOWS AND TO IMPLEMENT A FASTER   RAND0130
*          DIVISION SIMULATION ALGORITHM (SEE REFERENCE 2.)            RAND0140
*                                                                      RAND0150
*   CALLING SEQUENCE:                                                  RAND0160
*                                                                      RAND0170
*          1.) ONCE, AT THE BEGINNING OF THE PROGRAM, A CALL MUST BE   RAND0180
*              MADE TO OVERFLOW AS FOLLOWS:                            RAND0190
*              CALL OVFLOW                                             RAND0200
*              NO ARGUMENTS ARE REQUIRED FOR THIS CALL.                RAND0210
*                                                                      RAND0220
*          2.) TO RECEIVE RANDOM VARIATES, A CALL TO RANDOM IS MADE AS RAND0230
*              FOLLOWS:                                                RAND0240
*                        CALL RANDOM(IX,A,N)                           RAND0250
*              WHERE  IX  IS AN INTEGER*4 VARIABLE. PRIOR TO CALLING   RAND0260
*                         RANDOM FOR THIS FIRST TIME, IX SHOULD BE     RAND0270
*                         INITIALIZED TO ANY INTEGER VALUE IN THE RANGE RAND0280
*                         1 TO 2147483647. UPON RETURN FROM RANDOM IX IS RAND0290
*                         REPLACED BY THE NEXT INTEGER VARIATE IN THE  RAND0300
*                         SEQUENCE. IX SHOULD NOT BE CHANGED BY THE USER RAND0310
*                         SINCE IT WOULD ALTER THE PERFORMANCE OF RANDOM RAND0320
*                     A   IS A REAL*4 VARIABLE OR VECTOR WHICH WILL    RAND0330
*                         RECEIVE THE REAL RANDOM VARIATE(S). A NEED NOT RAND0340
*                         BE INITIALIZED AND MAY BE CHANGED BY THE USER RAND0350
*                     N   IS AN INTEGER*4 VARIABLE WHICH DETERMINES    RAND0360
*                         THE NUMBER OF REAL*4 VARIATES TO BE PLACED   RAND0370
*                         IN THE VECTOR A. IF ONLY INTEGER*4 VARIATES  RAND0380
*                         ARE DESIRED (I.E. THE RETURNED VALUE OF IX)  RAND0390
*                         THEN N MUST BE 1, OTHERWISE THE PROGRAM MAY  RAND0400
*                         ABEND.                                       RAND0410
```

```
*                                                                  RAND0420
*  EXAMPLES:                                                       RAND0430
*                                                                  RAND0440
*        1.) TO GENERATE 500 INTEGER*4 VARIATES:                   RAND0450
*                  DIMENSION  IRAND(500)                           RAND0460
*                  IRAND(1)=1912615462                             RAND0470
*                  N=1                                             RAND0480
*                  CALL OVFLOW                                     RAND0490
*                  DO 1 J=1,500                                    RAND0500
*                  CALL RANDOM(IRAND(J),AX,N)                      RAND0510
*                1 CONTINUE                                        RAND0520
*                     . . .                                        RAND0530
*                     . . .                                        RAND0540
*                                                                  RAND0550
*        2.) TO GENERATE 1000 REAL*4 VARIATES:                     RAND0560
*                  DIMENSION  RAND(1000)                           RAND0570
*                  CALL OVFLOW                                     RAND0580
*                  IA=284118599                                    RAND0590
*                  CALL RANDOM(IA,RAND,1000)                       RAND0600
*                     . . .                                        RAND0610
*                     . . .                                        RAND0620
*                                                                  RAND0630
*  METHOD: THE LEHMER MULTIPLICATIVE SCHEME IS USED. THE RECURRENCE     RAND0640
*          RELATION IS   X(N+1)= A*X(N) MOD(P)  WHERE P IS (2**31)-1.   RAND0650
*          AND A IS A POWER OF A POSITIVE PRIMITIVE ROOT OF P (7**5 =   RAND0660
*          16807).  THE PRECEEDING INTEGER*4 VARIATE IS MULTIPLIED      RAND0670
*          BY A AND IF THE RESULT IS LESS THAN P, IT IS RETURNED AS THE RAND0680
*          NEXT INTEGER*4 VARIATE.  IF THE RESULT IS GREATER THAN P,    RAND0690
*          A FIXED POINT OVERFLOW WILL OCCUR.  THE CALL TO OVFLOW HAS   RAND0700
*          ESTABLISHED THAT RANDOM WILL HANDLE THIS OVERFLOW CONDITION. RAND0710
*          RATHER THAN DIVIDE THE RESULT BY P, DIVISION IS SIMULATED    RAND0720
*          AND A VALID RESULT IS RETURNED. IN EITHER CASE THE RESULTING RAND0730
*          INTEGER*4 VARIATE IS CONVERTED TO A REAL*4 VARIATE AND THEN  RAND0740
*          NORMALIZED TO THE INTERVAL(0.,1.). IF MORE THAN ONE REAL*4   RAND0750
*          VARIATE IS REQUESTED, RANDOM LOOPS, FILLING SUCCESSIVE       RAND0760
*          ELEMENTS OF THE VECTOR A.                                    RAND0770
*                                                                       RAND0780
*  TIMING: WHEN CALLING RANDOM FOR MORE THAN ONE REAL*4 VARIATE AT A    RAND0790
*          TIME, THE OVERHEAD FOR THE CALLING SEQUENCE IS MINIMIZED.    RAND0800
*          THE AVERAGE TIME PER VARIATE IS  15.6 MICROSECONDS BASED ON  RAND0810
*          AN EXPERIMENT CALLING FOR 1000 VARIATES PER CALL FOR 1000    RAND0820
*          CALLS, OR A TOTAL OF 1,000,000 VARIATES.                     RAND0830
```

```
*                                                                        RAND0840
*   REFERENCES:                                                          RAND0850
*                                                                        RAND0860
*         1.) LEWIS,P.A.W.,A.S.GOODMAM,AND J.M.MILLER,'A PSEUDO-RANDOM   RAND0870
*              NUMBER GENERATOR FOR THE SYSTEM/360.', IBM SYSTEMS        RAND0880
*              JOURNAL,VOL.8,NUMBER 2,1969,PP.136-146.                   RAND0890
*                                                                        RAND0900
*         2.) PAYNE,W.H.,J.R.RABUNG,AND T.P.BOGYO,'CODING THE LEHMER     RAND0910
*              PSEUDO-RANDOM NUMBER GENERATOR,' COMMUNICATIONS OF THE    RAND0920
*              ACM, VOL.12,NUMBER 2, FEBRUARY 1969, PP.85-86.            RAND0930
*                                                                        RAND0940
*   CATEGORY V1: RANDOM NUMBER GENERATORS                                RAND0950
*                                                                        RAND0960
*   CHECKED OUT AT NPS ON IBM 360/67 BY G. LEARMONTH, APRIL 1972.        RAND0970
*                                                                        RAND0980
*                                                                        RAND0990
************************************************************************ RAND1000
         EJECT                                                           RAND1010
OVFLOW   CSECT                                                           RAND1020
         ENTRY RANDOM                                                    RAND1030
         USING OVFLOW,R12                                                RAND1040
         SAVE  (14,12),,*                                                RAND1050
         LR    R12,R15           GET ADDRESSIBILITY                      RAND1060
         ST    R13,SA+4          SAVE CALLER'S R13                       RAND1070
         LR    R2,R13                                                    RAND1080
         LA    R13,SA            POINT AT MY SAVE AREA                   RAND1090
         ST    R13,8(,R2)        POINT CALLER AT MY SAVE AREA            RAND1100
*                                                                        RAND1110
*   ISSUE SPIE MACRO TO FIELD INTERRUPTS                                 RAND1120
*                                                                        RAND1130
         SPIE  XIT,(8,9,11,12,13,15)                                     RAND1140
*                                                                        RAND1150
         ST    R1,PICA           SAVE PROGRAM INTERRUPT CONTROL ADDR.    RAND1160
         L     R13,SA+4          RESTORE CALLER'S R13                    RAND1170
         RETURN (14,12)          ALL DONE , GO BACK...                   RAND1180
*                                                                        RAND1190
*   HERE ON INTERRUPT                                                    RAND1200
*                                                                        RAND1210
         USING XIT,R15                                                   RAND1220
XIT      TM    7(R1),X'F7'       FIXED POINT OVERFLOW?                   RAND1230
         BC    5,FORT            NO, GIVE CONTROL TO OS TO HANDLE        RAND1240
         CLC   17(3,R1),ARANDOM+1 DOES BASE POINT TO RANDOM?             RAND1250
         BNE   0(,R14)           IF NOT, IGNORE THE INTERRUPT            RAND1260
```

```
*                                                                       RAND1270
*   MAKE THE CORRECTION, I.E. ADD 2**31-3 TO R4                         RAND1280
*                                                                       RAND1290
         A     R4,PM2                                                   RAND1300
         AR    R4,R2                                                    RAND1310
         BR    R14              DONE, RETURN TO INTERRUPT HANDLER       RAND1320
*                                                                       RAND1330
*   LET OS HANDLE THE INTERRUPT                                         RAND1340
*                                                                       RAND1350
FORT     L     R15,PICA         LOAD R15 WITH OS ROUTINE ADDRESS        RAND1360
         L     R15,0(,R15)      CLEAR HIGH ORDER BYTE                   RAND1370
         BR    R15              GO...                                   RAND1380
         EJECT                                                          RAND1390
*                                                                       RAND1400
*   GENERATE SOME RANDOM NUMBERS                                        RAND1410
*                                                                       RAND1420
         USING *,R15                                                    RAND1430
RANDOM   SAVE  (14,12),,*                                               RAND1440
         ST    R13,SA+4         SAVE CALLER'S R13                       RAND1450
         LR    R2,R13                                                   RAND1460
         LA    R13,SA           POINT AT MY SAVE AREA                   RAND1470
         ST    R13,8(,R2)       POINT CALLER AT MY SAVE AREA            RAND1480
         SUR   FR0,FR0          CLEAR FLOATING POINT REG. 0             RAND1490
         LM    R9,R11,A75       LOAD R9 WITH A=7**5                     RAND1500
*                               LOAD R10 WITH FLOATING POINT +1         RAND1510
*                               LOAD R11 WITH NORMALIZATION COMPARAND   RAND1520
         LA    R2,4             WORD INCREMENT FOR BXLE                 RAND1530
         LM    R5,R7,0(R1)      LOAD ADDRESSES OF THREE ARGUMENTS       RAND1540
         L     R5,0(,R5)        GET FIRST ARGUMENT, IX=SEED             RAND1550
         L     R3,0(,R7)        GET THIRD ARGUMENT, N=NUMBER DESIRED    RAND1560
         SLA   R3,2             CONVERT WORDS TO BYTES                  RAND1570
         SR    R6,R2            POINT TO WORD BEFORE X                  RAND1580
         LR    R7,R2                                                    RAND1590
         LA    R12,NONORM                                               RAND1600
         CNOP  2,8              ALIGN TO SECOND BYTE OF DOUBLE WORD     RAND1610
```

```
*                                                                     RAND1620
*   LOOP FOR N RANDOM NUMBERS                                         RAND1630
*                                                                     RAND1640
LOOP     MR    R4,R9             (R5=IX)*(R9=A)                       RAND1650
         SLDA  R4,1              DIVIDE BY P FOR REMAINDER            RAND1660
         SRL   R5,1              R4=REMAINDER; R5=QUOTIENT            RAND1670
         AR    R4,R5             Q+R = NEXT IX                        RAND1680
         LR    R5,R4             R5 = IX FOR NEXT GO AROUND           RAND1690
         SRL   R4,7              CONVERT R4 TO FLOATING POINT         RAND1700
         OR    R4,R10            OR ON AN EXPONENT FIELD              RAND1710
         ST    R4,0(R7,R6)       STORE IN X INDEXED BY R7             RAND1720
         CR    R4,R11            NEED NORMALIZATION?                  RAND1730
         BNL   0(,R12)           NO, BRANCH AROUND                    RAND1740
         LE    FR2,0(R7,R6)      X(R7) TO FLOATING POINT REG. 2       RAND1750
         AER   FR2,FR0           NORMALIZE BY ADDITION OF 0           RAND1760
         STE   FR2,0(R7,R6)      STORE BACK NORMALIZED                RAND1770
NONORM   BXLE  R7,R2,LOOP                                             RAND1780
*                                                                     RAND1790
         L     R4,0(,R1)         GET ADDRESS OF IX AGAIN              RAND1800
         ST    R5,0(,R4)         STORE LAST INTEGER INTO IX           RAND1810
         L     R13,SA+4          RESTORE CALLER'S R13                 RAND1820
         RETURN (14,12)                                               RAND1830
         EJECT                                                        RAND1840
*                                                                     RAND1850
*   CONSTANTS AND STORAGE                                             RAND1860
*                                                                     RAND1870
PICA     DC    F'0'                                                   RAND1880
ARANDOM  DC    V(RANDOM)                                              RAND1890
PM2      DC    F'2147483645'     P=2**31-1   PM2=2**31-3              RAND1900
A75      DC    F'16807'                                               RAND1910
         DC    X'40000001'                                            RAND1920
         DC    X'40100000'                                            RAND1930
SA       DS    18F                                                    RAND1940
R0       EQU   0                                                      RAND1950
R1       EQU   1                                                      RAND1960
R2       EQU   2                                                      RAND1970
R3       EQU   3                                                      RAND1980
R4       EQU   4                                                      RAND1990
R5       EQU   5                                                      RAND2000
R6       EQU   6                                                      RAND2010
R7       EQU   7                                                      RAND2020
R8       EQU   8                                                      RAND2030
R9       EQU   9                                                      RAND2040
R10      EQU   10                                                     RAND2050
R11      EQU   11                                                     RAND2060
R12      EQU   12                                                     RAND2070
```

```
R13        EQU   13                                                        RAND2080
R14        EQU   14                                                        RAND2090
R15        EQU   15                                                        RAND2100
FR0        EQU   0                                                         RAND2110
FR2        EQU   2                                                         RAND2120
           END                                                             RAND2130
                     JOB COMPLETE


IEF285I   SYS72334.T093206.SV000.BOX213$L.R0000001      SYSOUT
IEF285I   VOL SER NOS= SPOOL2.
IEF285I   SYS72334.T093206.SV000.BOX213$L.R0000002      SYSOUT
IEF285I   VOL SER NOS= SPOOL3.
IEF285I   SYS1.PROCLIB                                  KEPT
IEF285I   VOL SER NOS= MVTRES.
IEF285I   SYS72334.T093206.RV000.BOX213$L.S0000003      SYSIN
IEF285I   VOL SER NOS= SPOOL3.
IEF285I   SYS72334.T093206.RV000.BOX213$L.S0000003      DELETED
IEF285I   VOL SER NOS= SPOOL3.
IEF373I STEP /PRINT    / START 72334.0934
IEF374I STEP /PRINT    / STOP  72334.0935 CPU    0MIN 00.76SEC MAIN   54K LCS    0K
IEF375I  JOB /BOX213$L/ START 72334.0934
IEF376I  JOB /BOX213$L/ STOP  72334.0935 CPU    0MIN 00.76SEC
JOB PARAMETERS: CORE-    0K COMPILE &  60K GO, DISK TRKS- 120 SYSOUT &    0 SYSDA,
                  5 SEC. CPU TIME,CLASS=A, PRTY= 7.
```

APPENDIX I.5. SAMPLING TWO CORRELATED VARIABLES

(i) General scheme. If $\underline{x}$ and $\underline{y}$ should show a correlation coefficient ρ (and should have mean and variance μ_1, σ_1^2 and μ_2, σ_2^2) then use

$$\underline{y} = a\underline{x} + b + \underline{u} \tag{5.1}$$

with

$$a = \rho \frac{\sigma_2}{\sigma_1}, \qquad b = \mu_2 - \rho \frac{\sigma_2}{\sigma_1} \mu_1$$

$$\underline{u} = (1 - \rho^2)^{1/2} \sigma_2 \underline{z} \tag{5.2}$$

where $\underline{u}$ is independent of $\underline{x}$, and $\underline{z}$ has mean 0 and variance 1.

Proof:

$$E(\underline{y}) = aE(\underline{x}) + b + E(\underline{u}) = \left(\rho \frac{\sigma_2}{\sigma_1} \mu_1\right) + \left(\mu_2 - \rho \frac{\sigma_2}{\sigma_1} \mu_1\right) + 0 = \mu_2 \tag{5.3}$$

$$\text{var}(\underline{y}) = a^2\sigma_1^2 + \sigma_u^2 = \rho^2 \frac{\sigma_2^2}{\sigma_1^2} \sigma_1^2 + (1 - \rho^2)\, \sigma_2^2 = \sigma_2^2 \tag{5.4}$$

$$\begin{aligned} E(\underline{x}\,\underline{y}) &= E[a\underline{x}^2 + b\underline{x} + \underline{ux}] \\ &= \rho \frac{\sigma_2}{\sigma_1} (\sigma_1^2 + \mu_1^2) + \left(\mu_2 - \rho \frac{\sigma_2}{\sigma_1} \mu_1\right) \mu_1 + E(\underline{u})\, E(\underline{x}) \\ &= \rho\sigma_2\sigma_1 + \rho \frac{\sigma_2}{\sigma_1} \mu_1^2 + \mu_1\mu_2 - \rho \frac{\sigma_2}{\sigma_1} \mu_1^2 + 0 \\ &= \rho\sigma_2\sigma_1 + \mu_1\mu_2 \end{aligned} \tag{5.5}$$

$$\text{corr}(\underline{x},\underline{y}) = \frac{E(\underline{x}\,\underline{y}) - E(\underline{x})\, E(\underline{y})}{\sigma_1\sigma_2} = \frac{\rho\sigma_2\sigma_1 + \mu_1\mu_2 - \mu_1\mu_2}{\sigma_1\sigma_2} = \rho \tag{5.6}$$

(ii) A more stringent condition is that $\underline{x}$ and $\underline{y}$ should have a particular distribution type (besides satisfying the conditions on means and variances). Then the above scheme might still work if $\underline{u}$ is selected correctly. For example, suppose that $\underline{x}$ and $\underline{y}$ should be normal variates. The sum of normal variables remains normal.

Using (5.1) we take

$$\underline{y} = \rho \frac{\sigma_2}{\sigma_1} \underline{x} + \mu_2 - \rho \frac{\sigma_2}{\sigma_1} \mu_1 + (1 - \rho^2)^{1/2} \sigma_2 \underline{z} \qquad (5.7)$$

where $\underline{z}$ is $N(0,1)$. This special case of (5.1) corresponds with the result derived in another way in Naylor et al. (1967a, p. 99). For other distribution types no such simple schemes work. A possible (approximative) solution may follow from the next discussion.

(iii) Maximum dependence. It may be that the correlation, or more generally the dependence, is not meant to be fixed to a particular value but to be maximal.

(1) Maximum linear correlation: $\rho = +1$ or $\rho = -1$. Equation (5.1) yields

$$\underline{y} = \alpha \underline{x} + \beta \qquad (5.8)$$

with

$$\alpha = \frac{\sigma_2}{\sigma_1}, \quad \beta = \mu_2 - \frac{\sigma_2}{\sigma_1} \mu_1 \quad \text{for} \quad \rho = +1 \qquad (5.9)$$

and

$$\alpha = -\frac{\sigma_2}{\sigma_1}, \quad \beta = \mu_2 + \frac{\sigma_2}{\sigma_1} \mu_1 \quad \text{for} \quad \rho = -1 \qquad (5.10)$$

(2) Maximum positive dependence. We may use the same random numbers for generating $\underline{x}$ and $\underline{y}$.

(a) Continuous, increasing distributions for $\underline{x}$ and $\underline{y}$. If

$$\underline{x} = F_1^{-1}(\underline{r}) \qquad (5.11)$$

then

$$\underline{y} = F_2^{-1}(\underline{r}) = F_2^{-1}[F_1(\underline{x})] = g(\underline{x}) \qquad (5.12)$$

so that

$$\text{var}(\underline{y} | \underline{x} = x) = 0, \ E(\underline{y} | \underline{x} = x) = g(x) \qquad (5.13)$$

Note that $g(x)$ may be linear as in (5.8); e.g., for exponential $\underline{x}$ and $\underline{y}$ we have

$$y = -\frac{1}{\lambda_2}\,\ell n(r) = -\frac{1}{\lambda_2}\,\ell n(e^{-\lambda_1 x}) = \frac{\lambda_1}{\lambda_2}\,x \qquad (5.14)$$

(b) Discrete distribution. In Fig. 5.1 r_1 and r_2 yield the same x but different y. Hence $var(\underline{y}|\underline{x} = x) > 0$. So $\underline{y}$ is not completely dependent on $\underline{x}$. Anyhow, $\underline{x}$ and $\underline{y}$ may be expected to have positive correlation. If the distributions are symmetric so their mean and median are identical, then it is certain that $\underline{x}$ and $\underline{y}$ have positive correlation. (Compare: if $\underline{r} > 0.5$ then $\underline{x} > \mu_1$ and $\underline{y} > \mu_2$.)

(3) Maximum negative dependence. Use r and its complement $(1 - r)$ for $\underline{x}$ and $\underline{y}$, respectively. For $\underline{x}$ and $\underline{y}$ having the same distributions $(F_1 = F_2 = F)$, Andréasson (1971, p.5) proved that this procedure minimizes the correlation ρ; if F is symmetric then $\rho = -1$.

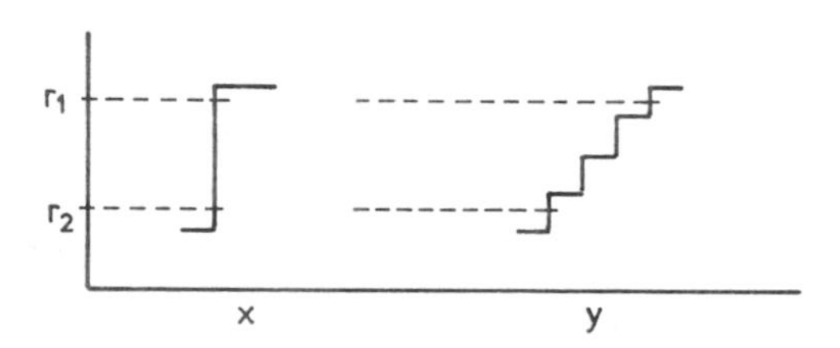

Fig. 5.1. Sampling $\underline{x}$ and $\underline{y}$ with common random numbers.

EXERCISES[9)]

1. Why can the cycle length of the multiplicative generator in (6) never be longer than m? Why can the cycle length in the following (so-called Fibonnacci) generator be longer than m:

$$x_i = x_{i-1} + x_{i-2} \pmod{m} \quad ?$$

2. How would you use the Monte Carlo method to estimate

$$\int_a^{\infty} \exp\left\{-\frac{1}{2}\left(\frac{x-b}{c}\right)^2\right\} / \ln x \, dx$$

for given a, b, and c?

3. Give the flowchart for the single-server FIFO queuing system in Fig. 4.

4. Give the analytical derivation for $p = P(\underline{x} < a)$; see (2.1) through (2.4). What is the value for p if $a = 83$?

5. Adjust the flowchart in Fig. 3.1 to simulate strategy γ defined at the end of Appendix I.3.

6. Consider $P(\underline{x} = x_i) = p_i$ $(i = 1, \ldots, n)$. What is the advantage of rearranging the x_i in order of increasing p_i when sampling x?

7. Consider the economic model $\underline{y}_t = \alpha + \beta x_t + \beta \underline{\epsilon}_t$ with $\underline{\epsilon}_t : N(0, \sigma^2)$. Suppose (x_t, y_t) for $t = 1, \ldots, n$ is given. From which distribution would you sample the disturbance term? Would you also sample the estimated parameter values, $\underline{\hat{\alpha}}$ and $\underline{\hat{\beta}}$?

NOTES

1. For example, Ackoff (1971), Emery (1969), Emshoff and Sisson (1971, pp. 250-251), Eykhoff et al. (1966), and van Dixhoorn and Lyesen (1968). A bibliography on systems theory may be obtained from the Netherlands Society of Systems Research (subcommittee on systems methodology, c.o. K. von Raesfeld, Kohnstamm Instituut, Keizersgracht 73, Amsterdam).

2. Another method for evaluating a statistical technique (e.g., simple least squares), applies this technique to part of the historical data. This yields estimated parameters. These estimated parameters are used to "predict" the historical data that were not utilized in the estimation of the parameters. The "predicted" historical data can be compared with the actual historical data; see Graver (1969).

3. Remember that "$y = g(z)$ is a function" implies that every z corresponds to not more than one value of y. If $F(x)$ is a continuous, nondecreasing (cumulative distribution) function or a discrete (distribution) function then the inversion technique still works. For an exact derivation we refer to Haitsma and Oosterhoff (1964, pp. 12-13); practically no special problems arise as it follows from the sampling procedure for discrete distribution functions in Naylor et al. (1967a, p. 115 .

4. We owe this solution to the late L. Lips of the Katholieke Hogeschool Tilburg.

5. Another procedure suggested to us by H. Lombaers (Technische Hogeschool Delft) which reduces the computer time is as follows. From (20) we see that $\underline{x} < v$ if $\underline{r} > \exp(-\lambda v)$. For such values of $\underline{r}$ we do not calculate $\underline{x}$ since (1.4) shows that $g(\underline{x}) = 0$ anyhow. In this way we avoid the time-consuming log-calculation. So ξ is estimated by

$$\hat{\underline{\xi}} = \frac{-\lambda}{n} \sum_{j} \frac{1}{\ell n(\underline{r}_j)}$$

where the $\underline{r}_j$'s are those random numbers among the n random numbers not larger than the constant $\exp(-\lambda v)$.

6. In (1.1) λ must be positive since it is the parameter of the exponential density function. In Kleijnen (1968, p. 180) it is shown that v must also be positive for otherwise the integral in (1.1) would diverge.

7. One could argue that the variables in the flowchart should be underlined to indicate their stochastic character. Actually

these variables are symbolic computer memory addresses. Hence we do not underline variables in flowcharts.

8. The listing of this generator is reproduced with the kind permission of P. A. W. Lewis.

9. Many more exercises can be found in textbooks, e.g., Meier et al. (1969), Mihram (1972), Naylor et al. (1967a), and Schmidt and Taylor (1970).

REFERENCES

Ackoff, R.L. (1971). "Towards a system of systems concepts," Management Sci., 17, 661-671.

Ahrens, J.H. and U. Dieter (1972). "Computer methods for sampling from the exponential and normal distributions," Commun. ACM, 15, 873-882.

Andréasson, I.J. (1971). "On the generation of negatively correlated random numbers," Report Na 71.32, Department of Information Processing, The Royal Institute of Technology, Stockholm.

Andrews, D.F., P.J. Bickel, F.R. Hampel, P.J. Huber, W.H. Rogers, and J.W. Tukey (1972). Robust Estimates of Location. Princeton University Press, Princeton, N.J.

Bauknecht, K. (1971). "Programmiersprachen für Simulationen auf Rechenautomaten" (Programming languages for simulations on computers), in Digitale Simulation (K. Bauknecht and W. Nef, eds.), Springer, Berlin.

Benecke, R.W. and T.P. Westgaard (1972). On the Use of Risk Analysis in Business, University of Nebraska, Omaha, Neb.

Bovet, D.P. (1972). "On the Use of Models Employing Both Simulation and Analytical Solutions for Scheduling Computing Centers," in Working Papers, Vol. 2: Symposium Computer Simulation versus Analytical Solutions for Business and Economic Models, Graduate School of Business Administration, Gothenburg, Sweden.

Burt, J.M. (1972). Resource Allocation in Stochastic Project Net-Works: A Simulation and Programming Approach. Graduate School of Management, University of California, Los Angeles, Calif.

Burt, J.M., D.P. Graver, and M. Perlas (1970). "Simple stochastic networks: some problems and procedures," Naval Res. Logistics Quart., 17, 439-459.

Carter, E.E. (1971). "A simulation approach to investment decision," California Management Rev., 13, 18-26.

Chambers, J.M. (1970). "Computers in statistical research: simulation and computer-aided mathematics," Technometrics, 12, 1-15.

Chambers, R.P. (1967). "Random-number generation," IEEE Spectrum, 4, 48-56.

Churchman, C.W., R.L. Ackoff, and E.L. Arnoff (1959). Introduction to Operations Research, Wiley, New York.

Clough, D.J., J.B. Levine, G. Mowbray, and J.R. Walter (1965). "A simulation model for subsidy policy determination in the Canadian uranium mining industry," Can. Operational Res. Soc. J., 3, 115-128.

Cohen, K.J. (1960). Computer Models of the Shoe, Leather, Hide Sequence. Prentice-Hall, Englewood Cliffs, N.J.

Computing Reviews, Association for Computing Machinery, New York.

Conway, R.W. (1963). "Some tactical problems in digital simulation," Management Sci., 10, 47-61.

Conway, R.W., B.M. Johnson, and W.L. Maxwell (1959). "Some problems of digital systems simulation," Management Sci., 6, 92-110.

Coplin, W.D. (ed.) (1968). Simulation in the Study of Politics, Markham, Chicago, Ill.

Coveyou, R.R. and R.D. MacPherson (1967). "Fourier analysis of uniform random number generators," J. Assoc. Computing Machinery, 14, 100-119.

Dewan, P.B., C.E. Donaghey, and J.B. Wyatt (1972). "OSSL--A specialized language for simulating computer systems," in Proc. AFIPS 1972 Spring Joint Computer Conf., AFIPS Press, Montvale, N.J.

Donaldson, T.S. (1966). Power of the F-Test for Nonnormal Distribution and Unequal Error Variances, RM-5072-PR, The Rand Corporation, Santa Monica, Calif.

Driehuis, W. (1972). Fluctuations and Growth in a Near Full Employment Economy. Universitaire Pers, Rotterdam.

Dutton, J.M. and W.H. Starbuck (1971a). Computer Simulation of Human Behavior, Wiley, New York.

Dutton, J.M. and W.H. Starbuck (1971b). "Computer simulation models of human behavior: A history of an intellectual technology," IEEE Trans.:Systems, Man, Cybernetics, SMC-1, 128-171.

Elton, M. and J. Rosenhead (1971). "Micro-simulation of markets," Operational Res. Quart., 22, 117-144.

Emery, F.E. (ed.) (1969). Systems Thinking, Penguin, Harmondsworth, England.

Emshoff, J.R. and R.L. Sisson (1971). Design and Use of Computer Simulation Models, Macmillan, New York.

Eykhoff, P., P.M.E.M. van der Grinten, H. Kwakernaak, and B.P.T. Veltman (1966). "Systems modelling and identification," in Automatic and Remote Control, III, Proc. 3rd Congr. IFAC, United Kingdom Automation Council, London, 1966.

Fisz, M. (1967). Probability Theory and Mathematical Statistics, 3rd ed., Wiley, New York.

Flagle, C.D. (1960). "Simulation techniques," in Operations Research and Systems Engineering (C.D. Flagle, W.H. Huggins, and R.H. Roy, eds.), The Johns Hopkins Press, Baltimore, Md.

Forrester, J. (1961). Industrial Dynamics, M.I.T. Press, Cambridge, Mass.

Geisler, M.A. (1960). "The use of man-machine simulation for support planning," Naval Res. Logistics Quart., 7, 421-428.

Graver, C.A. (1969). Historical Simulation: A Procedure for the Evaluation of Estimating Procedures, Vol. 1, Contract DACH15-68-C-0364, General Research Corporation, Santa Barbara, Calif.

Guetzkow, H. (1971). "Simulations in the consolidation and utilization of knowledge about international relations," in Cybernetics, Simulation and Conflict Resolution (D.E. Knight, H.W. Curtis, and L.J. Fogel, eds.), Spartan Books, New York.

Guetzkow, H., P. Kotler, and R.L. Schultz (1972). Simulation in Social and Administrative Science, Prentice-Hall, Englewood Cliffs, N.J.

Haitsma, A.H. and J. Oosterhoff (1964). Monte Carlo Methoden (Monte Carlo methods), Rapport S 265 (C13), Leergang Besliskunde, Hoofdstuk XVI, Stichting Mathematisch Centrum, Amsterdam.

Halton, J.H. (1970). "A retrospective and prospective survey of the Monte Carlo method," SIAM Rev., 12, 1-63.

Hammersley, J.M. and D.C. Handscomb (1964). Monte Carlo Methods, Wiley, New York; Methuen, London.

Hanken, A.F.G. and B.G.F. Buys (1971). "Systems analysis and business models," Ann. Systems Res., 1, 9-16.

Hauser, N., N.N. Barish, and S. Ehrenfeld (1966). "Design problems in a process control simulation," J. Ind. Eng., 17, 79-86.

Heuts, R.M.J. and P.J. Rens (1972). A Monte Carlo Study of the Kuyper Test Statistic for Testing Exponentiality (Two Different Approaches), R.C.-notitie nr. 13, Rekencentrum Katholieke Hogeschool, Tilburg, Netherlands.

Hillier, F.S. and G.J. Lieberman (1968). Introduction to Operations Research, Holden-Day, San Francisco, Calif., Chapter 14.

Horn, W. (1968). Het Aanbod van Varkens in Nederland (The supply of hogs in the Netherlands), Centrum voor Landbouwpublicaties en Landbouwdocumentatie, Wageningen, Netherlands.

Howrey, E.P. (1972). "Selection and Evaluation of Econometric Models," in Working Papers, Vol. 4, Symposium Computer Simulation versus Analytical Solutions for Business and Economic Models, Graduate School of Business Administration, Gothenburg, Sweden.

Howrey, E.P. (1966). Stabilization Policy in Linear Stochastic Systems, Econometric Research Program, Princeton University, Princeton, N.J.

Howrey, P. and H.H. Kelejian (1969). "Simulation versus analytical solutions," in The Design of Computer Simulation Experiments, T.H. Naylor (ed.), Duke University Press, Durham, N.C. (Reprinted in: T.H. Naylor, Computer Simulation Experiments with Models of Economic Systems, Wiley, New York, 1971.)

Hutter, R. (1970). "Simulation eines Informationssystems mit Benutzerprioritäten" (Simulation of an information processing system with users' priorities), Ablauf Planungsforsch., 11, 238-243.

International Abstracts in Operations Research. International Federation of Operations Research Societies (c/o N. Miller, 428 E. Preston St., Baltimore, Md.).

Jansson, B. (1966). Random Number Generators, Victor Pettersons Bokindustri Aktiebolag, Stockholm.

Jessop, W.N. (1956). "Monte Carlo methods and industrial problems," Appl. Stat., 5, 158-165.

Johnston, J. (1963). Econometric Methods, McGraw-Hill, New York.

Kay, I.M. (1972). "An over-the-shoulder look at discrete simulation languages," in Proc. AFIPS 1972 Spring Joint Computer Conf., AFIPS Press, Montvale, N.J.

Kiviat, P.J. (1967). Digital Computer Simulation: Modeling Concepts, RM-5378-PR, The Rand Corporation, Santa Monica, Calif.

Kleijnen, J.P.C. (1970). Simulation Terminology and Application Areas, Working Paper, Katholieke Hogeschool, Tilburg, Netherlands.

Kleijnen, J.P.C. (1968). "Een toepassing van 'importance sampling'" (An application of importance sampling), Stat. Neerl., 22, 179-198. (Also published as report EIT, no. 2, of the Economisch Institut Tilburg, Econometrische Afdeling, Tilburg, Netherlands.)

Klingel, A.R. (1966). "Bias in PERT project completion time calculations for a real network," Management Sci., 13, 194-201.

Kmenta, J. and R.F. Gilbert (1968). "Small sample properties of alternative estimators of seemingly unrelated regressions," J. Amer. Stat. Assoc., 13, 1180-1200.

Knuth, D.E. (1969). The Art of Computer Programming, Vol. 2, Semi-Numerical Algorithms, Addison-Wesley, Reading, Pa.

Kohlas, J. (1971). "Die Monte Carlo Methode" (The Monte Carlo method), in Digitale Simulation (K. Bauknecht and W. Nef, eds.), Springer, Berlin.

Koopman, B.O. (1970). "A study on the logical basis of combat simulation," Operations Res., 18, 855-882.

Kriens, J. (1964). Markovketens (Markov chains), Katholieke Hogeschool, Tilburg, Netherlands.

Kriens, J. and G. de Leve (1966). Inleiding tot de Mathematische Besliskunde (Introduction to operations research), Leergang Besliskunde, I.5, Mathematisch Centrum, Amsterdam.

Lewis, P.A.W. (1972). Large-Scale Computer-Aided Statistical Mathematics, Naval Postgraduate School, Monterey, Calif., in Proc. Computer Science and Statistics: 6th Annual Symp. Interface, Western Periodical Co., Hollywood, Calif.

Lewis, P.A.W., A.S. Goodman, and J.M. Miller (1969). "A pseudo-random number generator for the system/360," IBM Systems J., 8, 136-147.

Lucas, H.C. (1971). "Performance evaluation and monitoring," Computing Surveys, 3, 79-91.

McCracken, D.D. (1955). "The Monte Carlo method," Scientific American, 192, 90-97.

Maguire, J.N. (1972). "Discrete computer simulation-technology and applications--the next ten years," Proc. AFIPS 1972 Spring Joint Computer Conf., AFIPS Press, Montvale, N.J.

Maisel, H. and G. Gnugnoli (1972). Simulation of Discrete Stochastic Systems, Science Research Associates, Palo Alto, Calif.

Mann, N.R. (1970). "Computer-aided selection of prior distributions for generating Monte-Carlo confidence bounds on system reliability," Naval Res. Logistics Quart., 17, 41-54.

Marsaglia, G. (1972). "The structure of linear congruential sequences," in Applications of Number Theory to Numerical Analysis, S.K. Zaremba (ed.), Academic, New York.

Marsaglia, G. and T.A. Bray (1968). "One-line random number generators and their use in combinations," Commun. ACM, 11, 757-759.

Marsaglia, G., K. Ananthanarayanan, and N. Paul (ca. 1972). How to Use the McGill Random Number Package "SUPER-DUPER," School of Computer Science, McGill University, Montreal.

Mededelingen Operationele Research (Information on Operations Research), Vereniging voor Statistiek (c/o J. Meinardi, Ant. Duyckstraat 102, The Hague).

Meier, R.C., W.T. Newell, and H.L. Pazer (1969). Simulation in Business and Economics, Prentice-Hall, Englewood Cliffs, N.J.

Mihram, G.A. (1972). Simulation: Statistical Foundations and Methodology, Academic, New York.

Mihram, G.A. (1971). "A glossary of simulation terminology," J. Stat. Computation Simulation, 1, 35-44.

Mikhail, W.M. (1972). "Simulating the small-sample properties of econometric estimators,: J. Amer. Stat. Assoc., 67, 620-624.

Molenaar, W. (1968). "Simulatie" (Simulation), in: Minimaxmethode, Netwerk Planning, Simulatie (J. Kriens, F. Gobel, and W. Molenaar, eds.) Leergang Besliskunde, I.8, Mathematisch Centrum, Amsterdam.

Moore, J.M. and R.C. Wilson (1967). "A review of simulation research in job shop scheduling," Production Inventory Management, 8, 1-10.

Morgenthaler, G.W. (1961). "The theory and application of simulation in operations research," in Progress in Operations Research (R.L. Ackoff, ed.), Wiley, New York.

Morse, P.M. (1962). "Markov processes," in Notes on Operations Research, assembled by Operations Research Center, M.I.T., The M.I.T. Press, Cambridge, Mass.

Nance, R.E. and C. Overstreet (1972). "A bibliography on random number generation," Computing Rev., 13, 495-508.

Naylor, T.H. (1971). Computer Simulation Experiments with Models of Economic Systems, Wiley, New York.

Naylor, T.H. (1969). "Bibliography 19; Simulation and gaming," Computing Rev., 10, 61-69.

Naylor, T.H., J.L. Balintfy, D.S. Burdick, and K. Chu (1967a). Computer Simulation Techniques, Wiley, New York.

Naylor, T.H., D.S. Burdick, and W.E. Sasser (1967b). "Computer simulation experiments with economic systems: The problem of experimental design," J. Amer. Stat. Assoc, 62, 1315-1337.

Naylor, T.H., K. Wertz, and T. Wonnacott (1968). "Some methods for evaluating the effects of economic policies using simulation experiments," Rev. Intern. Stat. Inst., 36, 184-200.

Neave, H.R. and C.W.J. Granger (1968). "A Monte Carlo study comparing various two-sample tests for differences in mean," Technometrics, 10, 509-522.

Neeleman, D. (1973). Multicollinearity in Linear Economic Models, Tilburg University Press, Tilburg, Netherlands.

Nelson, R.T. (ca. 1966). Systems, Models and Simulation. Mimeographed notes, Graduate School of Business Administration, University of California at Los Angeles, Los Angeles, Calif.

Netherlands A.D.P. Research Center (1967). Simulation, Series special bibliographies no. SB 39, Amsterdam.

Newman, T.G. and P.L. Odell (1971). The Generation of Random Variates, Griffin, London.

Niedereichholz, J. (1971). "Business simulation and management decisions," Management Intern. Rev., 11, 47-52.

Nisenfeld, A.E. (1971). "Analog and digital simulation: A comparison," Instrument Control Systems, 44, 91-93.

Operations Research-Management Science, Executive Sciences Institute, Inc., Whippany, N.J.

Orcutt, G.H., M. Greenberger, J. Korbel, and A.M. Rivlin (1961). Microanalysis of Socioeconomic Systems: A Simulation Study, Harper and Row, New York.

Pomerantz, A.G. (1970). "Predict your system's fortune: use simulation's crystal ball," Computer Decisions, 2, 16-19.

Rechtschaffen, R.N. (1972). "Queuing simulation using a random number generator," IBM Systems J., 11, 255-271.

Reith, P.F. (1969). "Computers in de econometrie" (Computers in econometrics), IBM Kwartaalschrift, 6, 49-53.

Rose, J. (1970). Progress of Cybernetics, Vols. 1, 2, 3, Gordon and Breach, London.

Rytz, R. (1971). "Sim--ein neues Simulationskonzept" (SIM--A new simulation concept), in Digitale Simulation (K. Bauknecht and W. Nef, eds.), Springer, Berlin.

Salazar, R.C. and S.K. Sen (1968). "A simulation model of capital budgeting under uncertainty," Management Sci., Application Ser., 15, 161-179.

Sasser, W.E. (1969). A Causal Relationship between a Model's Characteristics and the Performances of the Estimators of the Model's Parameters: A Pilot Study, Graduate School of Business Administration, Harvard University, Boston, Mass.

Schink, W.A., and J.S.Y. Chiu (1966). "A simulation study of effects of multi-collinearity and autocorrelation on estimates of parameters," J. Financial Quantitative Analysis, 1, 36-67.

Schmidt, J.W. and R.E. Taylor (1970). Simulation and Analysis of Industrial Systems, Richard D. Irwin, Inc., Homewood.

Shannon, R.E. and W.E. Biles (1970). "The utility of certain curric-culum topics to operations research practitioners," Operations Res., 18, 741-745.

Shreider, Y.A. (ed.) (1964). Method of Statistical Testing: Monte Carlo Method, Elsevier, Amsterdam.

Shubik, M. (1972). On gaming and game theory. Management Sci., Professional Series, 18, 37-53.

Shubik, M. (1970). A Preliminary Bibliography on Gaming, Department of Administrative Sciences, Yale University, New Haven. Conn.

Shubik, M. (1960). "Bibliography on simulation, gaming, artificial intelligence and allied topics," J. Amer. Stat. Assoc., 55, 736-751.

Shubik, M. and G. Wolf (1972). A List of Publications Relevant to the Game and Its Analysis, Appendix B of Gaming Newsletter, Depart-ment of Administrative Sciences, Yale University, New Haven. Conn.

Sisson, R.L. (1969). "Simulation: Uses," in Progress in Operations Research, Vol. 3 (Y.S. Aronofsky, ed.), Wiley, New York.

Smith, J. (1968). Computer Simulation Models, Griffin, London.

Smith, J.U.M. (1970). "Computer simulation of industrial operations," in Progress of Cybernetics, Vol. 2 (J. Rose, ed.), Gordon and Breach, London.

Teichroew, D. (1965). "A history of distribution sampling prior to the era of the computer and its relevance to simulation," J. Amer. Stat. Assoc., 60, 27-49.

Ten Broeke, A.M. (1972). "Commentaar op artikel Simulating inter-arrival times with time-dependent arrival rates" (Comment on article Simulating inter-arrival times with time-dependent arrival rates), Mededelingen Operationele Res., 11, 174-177.

Theil, H. and J.C.G. Boot (1962). "The final form of econometric equation systems," Rev. Intern. Stat. Inst., 30, 136-152.

Tocher, K.D. (1963). The Art of Simulation, English Universities Press, London.

Tocher, K.D. (1966). "The state of the art of simulation--A survey," Proc. 4th Intern. Conf. Operational Res.

Turban, E. (1972). "A sample survey of operations-research activities at the corporate level," Operations Res., 20, 708-721.

Van Daal, J. (1972). "Antwoord" (Reply), Mededelingen Operationele Res., 11, 208.

Van Daal, J. and F. van Doeland (1972). Waiting for the Bridge, Report 7208, Econometric Institute, Netherlands School of Economics, Rotterdam.

Van Dixhoorn, J.J. and D.P. Lyesen (1968). Simulatie: Methodologische en Systeemtheoretische Aspecten. (Simulation: methodological and systems-theoretic aspects), Kenmerk 2410330, Dhn/Sks, Afdeling der Electrotechniek, Technische Hogeschool Twente, Enschede, Netherlands.

Van Doeland, F. and J. van Daal (1972). "Simulating inter-arrival times with time-dependent arrival rates," Mededelingen Operationele Res., 11, 107-119.

Van Slyke, R.M. (1963). "Monte Carlo methods and the PERT problem," Operations Res., 11, 839-860.

Chapter II

THE STATISTICAL ASPECTS OF SIMULATION[1)]

II.1. INTRODUCTION AND SUMMARY

In this chapter we shall give a survey of the statistical aspects of simulation. Emphasis will be on input data analysis, starting conditions, variance reduction techniques, validation, experimental designs (including screening designs and response surface methodology), multiple ranking and comparison procedures, sample size in terminating and nonterminating systems, and runlength in steady-state simulations. It would be impossible to discuss all statistical aspects of simulation in complete detail. Therefore we shall concentrate on a few topics in later chapters. In the present chapter we shall give references as complete as possible to enable the reader to study the other topics on his own.

As we saw in Chapter I we define simulation as experimenting with an abstract model over time. Most simulation models contain stochastic variables, e.g., stochastic service times. We shall restrict our attention to these stochastic or "Monte Carlo" simulations. Stochastic simulation is actually a sampling experiment (although it is a special type of experiment as it is performed on a model instead of a real-life object). Hence by its very nature simulation has many statistical aspects.

In any statistical experiment a careful design and analysis are necessary. A statistical analysis of the experiment is required (i) to extract all information from the experiment that is contained in it, and (ii) to reveal the limitations of the conclusions that are based on statistical data. However, we cannot obtain more information from an experiment than is contained in it. So it is the task of the statistical design to provide as much relevant information, for given cost, as possible. Though simulation is a statistical sampling experiment the simulation technique is usually not applied by statisticians but by engineers, social scientists, etc. (since the technique also requires model building). Consequently, many users of simulation have no specialized knowledge of mathematical statistics. We think, moreover, that the experimenters tend to spend most of their time on the development of the simulation model and the subsequent construction of the computer program. An exception to the rule is the fire department simulation by Carter and Ignall (1970, p. 18) who spent between 1/2 and 1 man-year on the construction of the model, but more time on the design and analysis. [The importance of developing statistical design and analysis techniques for simulation is also stressed by Wagner (1971, p. 1274).] It is our hope that this chapter will make the reader aware of the statistical aspects of simulation and will provide him with references for further study. Unfortunately, the investigation of these aspects started only recently, so some conclusions and recommendations are provisional and require more research.

Let us have a closer look at the use of simulation by social scientists. Until the application of simulation social scientists could in general not perform controlled experiments. Now simulation has made possible at least some form of controlled experimentation. This, however, confronts these scientists with a new kind of problem, namely, the statistical design and analysis of such experiments. For, unlike students in the engineering sciences, most students in the social sciences receive little instruction in the techniques for the design of experiments and the concomitant analysis, since it is assumed that they will not need these techniques. Therefore, the social scientists are in a rather unfavorable position [see Naylor, et al. (1967b, pp. 1315-1316) and Naylor et al. (1968, pp. 184-185)]. (The latter authors compare the various kinds of data an economist may use, namely, cross-section data, real-life experimental data, and model experimentation data.) We shall discuss the statistical aspects of simulation following the various steps in the development of a computer simulation as described in Naylor et al (1967a, pp. 23-41). We refer to the literature[2)] for a description of the planning process itself that starts with the initial formulation of the problem to be simulated, and ends with the ultimate analysis of the simulation results, with many feedbacks in between. We shall discuss only the statistical aspects of the various steps in this process, and we shall combine and rearrange the various steps given by Naylor et al. (1967a).

II.2. FORMULATION OF THE PROBLEM

Naylor et al. (1967a, p. 26) list as step 1 in the planning of a simulation experiment the formulation of the problem. This step has two statistical aspects.

(i) Definition of the output. We have to define the output variable or variables that are of interest to us, i.e., what do we want to measure? Fishman and Kiviat (1967, p. 24) distinguish two simulation purposes: the comparison of responses for different operating rules, and the determination of the functional relationships between the response and the input factors. Naylor et al. (1967a, p. 338) and Naylor et al. (1967b, p. 1333) distinguish as purposes the determination of the optimum combination of factors and the general investigation of the relationship between the response and the factors. Obviously the aims given by Fishman and Kiviat and by Naylor et al. do not conflict with each other. We shall see later that special statistical techniques are available for certain purposes of a simulation study (e.g., RSM for determining the optimum factor combination). Overholt (1968, pp. 64, 66) points out that in large-scale simulations the output of the model may be of interest to several levels of management, each level with its own interest. In that case it is difficult to determine the actual aim of the simulation.

(ii) Statistical reliability. We have to specify the statistical "accuracy" with which the output should be observed. In hypothesis testing we may require that the probability of erroneously rejecting a true hypothesis be α and the probability of erroneously accepting (or more precisely, nonrejecting) a false hypothesis be β. In estimation problems we may demand that the sample mean $\underline{\bar{x}}$ differ no more than δ units from the population mean μ with probability $(1 - \alpha)$. (Remember that stochastic variables like the sample mean are underlined in this book.)

Notice that the above formulation of statistical reliability demonstrates that we shall use classical statistical theory and no decision theory. For in the latter theory no fixed statistical reliability measures like the α- and β-error are specified. Instead loss functions and, often, prior information (so called Bayesian approach) are used; see, e.g., Schlaifer (1959) or the resumé in Wetherill (1966, pp. 85-110). We further remark that most of the time we shall apply parametric statistical techniques.

II.3. INPUT DATA ANALYSIS

Step 2 in the planning of a simulation experiment is the "collection and processing of real world data"; see Naylor et al.(1967a, p. 27) and Emshoff and Sisson (1971, pp. 55-56). The analysis of the input data may suggest that service time is exponentially distributed. After we have hypothesized a certain form of the distribution of the input variable (using input data or theoretical considerations) we can apply a goodness-of-fit test to check if the real-world data agree with the assumed theoretical form. Goodness-of-fit tests are discussed by Fishman and Kiviat (1967, pp. 20-21) and Naylor and Finger (1967, p. 98). A simulation application can be found in Pegels (1969, p. 223). Besides specifying the form of the functional relationships we have to estimate the values of the parameters of the relationships. Elaborate examples are given by Schmidt and Taylor (1970, pp. 481-498).

In some simulations (e.g., critical path calculations, capital budgeting) the input data comprise <u>subjective</u> probability estimates (obtained from managers, etc.). Kotler (1970) discusses procedures for acquiring these subjective estimates. Note that Chan (1971) uses the entropy criterion to derive the probability distribution in case we do not know anything about the input except its range.

It may be that <u>several forms</u> of the distribution of the input variable are accepted (or better, not rejected) by a goodness-of-fit test, especially if only a small amount of input data is available. A <u>sensitivity</u> analysis can then be performed to see if the output is sensitive to variations in the form of the input distributions. If the output is sensitive, then more information on the input should be gathered. A useful reference is Chambers and Mullick (1968, pp. 105-107); also see Clough et al. (1965, pp. 127-128) and Marks (1964, pp. 255-256). We add that it may very well be that no additional information can be obtained. In that case we might follow a procedure applied by Cohen (1960, p. 36). He determined some parameter values by simulating the model for various values and selecting that value which minimizes the sum of the squared deviations between the actual and the simulation output; also consult Emshoff and Sisson (1971, pp. 203, 216, 278).

Naylor et al. (1967a, pp. 34, 36) observe that besides estimating the values of the parameters the statistical <u>significance</u> of the estimates should be tested. Moreover, if a regression model is used the underlying assumptions should be tested to see if they are not violated by heteroscedasticity, autocorrelation, etc. We point out that such testing is indeed common in econometric regression models. A whole body of estimation and testing techniques for such models can be found in Johnston (1963). However, as we mentioned in Section I.6.2 a typical simulation model is not based on regression equations, but on a flowchart describing the relationships among the various system components, these relationships being based on prior theoretical knowledge. In a regression model we relate all variables with each other in a linear equation (or set of equations) and we test if a particular variable should indeed be included in that equation, i.e., we test if the estimated parameter is significantly different from zero. But let us consider the simulation model of a simple queuing problem. A parameter of interest is the parameter of the exponential distribution that is assumed to hold for the service times. This parameter must be positive, so we do not test whether the estimated parameter does differ significantly from zero!

We saw how actual data can be used to specify a model. Instead of using the historical data only for the specification of a theoretical distribution, we can use the data themselves as input to the model.[3] Therefore we distinguish the following possibilities.

(i) The actual data are used only to formulate a theoretical distribution. Sampling values from this distribution is performed by an appropriate transformation of random numbers, as we saw in Section I.6.4.

(ii) The actual data provide an empirical distribution from which we sample input data for the simulation experiment. Sampling is again performed by means of random numbers.

(iii) The historical data are used in the simulation in the order in which they were realized in practice. So instead of internal generation of input data by means of random numbers, the input data are now read into the computer.

Let us compare these three possibilities. Reading in historical data as in (iii) we recommend for the "validation" of the model. When validating a model we read in the historical data, have these data processed by the simulation program, obtain simulation output, compare this simulation output with the historical output, and decide whether the model is realistic or not. We shall return to validation in Section II.5. Once we have validated the model we shall use it to predict the response for certain system variants. Then we are no longer interested in "reproduction of the past." The historical data in (ii) and (iii) are no more than a sample fr[illegible] [illegible]ation of data. The theoretical distribution is supposed t[illegible] adequately this population. So after the validation of the model we sample from the population represented by the theoretical distribution as in (i).

II.4. THE MODEL AND COMPUTER PROGRAM

Step 3 in the planning of the simulation experiment is the formulation of a model. Parts of the model have already been formulated in Step 2 (Section II.3) where we specified certain distributions of the input variables. Obviously we still have some degree of freedom left when specifying the rest of the model. So the real-world system can be represented by various models. If several models pass the validation tests the simplest model should be used. A simple model may be one with recursive instead of simultaneous relations. For, as Naylor et al. (1967a, p. 33) observe, a recursive model saves computer time and simplifies the statistical estimation of the parameters.

Once we have formulated the model we can proceed to the computer program. A typical aspect of simulation is the manipulation of the model over time. Consequently, we have to specify the starting conditions of the simulated system. We shall distinguish between two situations: steady-state and transient-state simulations.

(i) Steady-state simulations. Many simulations have been run for the investigation of the steady-state performance of the system. We call a system in its steady state if the probability of being in one of its states is governed by a fixed probability function; otherwise the system is in the transient state. (Notice that in the steady, stationary, or equilibrium "state" or "condition" the system can still move from one "state" to another, e.g., the queuelength can change from the state "3 customers waiting" to the state "4 customers waiting.") These fixed probabilities are the limits of the probabilities that are realized after a long period of time and they are supposed to be independent of the state in which the system started. So the steady-state characteristics are by definition independent of the starting conditions of the simulation experiment.[4)]

We could take any starting conditions insofar as any initial conditions will lead to the same steady state. However, it is still relevant to select "good" starting conditions. For as Nelson (1966, pp. 14-15) points out the steady state is a limiting case which in an actual simulation experiment is never exactly realized, so the actual initial conditions bias the result. Nevertheless, we may hope that the simulation run will be so long as to make this bias of no practical importance. Of more practical interest is the following aspect. If we are only interested in steady-state characteristics then we usually throw away the observations made in the transient state, i.e., the state passed between the starting of the simulation and the arrival at the steady state. (However, later we shall discuss the Crane-Iglehart procedure that uses all observations.) This implies that we have to determine when the transient state is over and the steady state starts. If we overestimate the length of the transient state we throw away information on the steady state and this increases the variance of the estimator. (Computer time is wasted.) On the other hand, if we underestimate the length of the transient state then transient-state observations are treated as steady-state observations, and we bias the estimator of the steady-state response. We do not know a statistical technique for determining when the transient state ends (Conway et al., 1959, p. 108). Rules of thumb are given by Conway (1963, p. 49), Emshoff and Sisson (1971, p. 192), and Tocher (1963, p. 1746). Following Conway we may throw away the initial observations as long as they seem to increase steadily (or decrease, depending on the problem).[5)] He advises not to use the cumulated observations since a cumulative statistic lags behind and therefore implies that the transient state is overestimated. Nevertheless, in practice the cumulated observations are often used [cf. Gürtler (1969, pp. 94-96)]. Van Daal and van Doeland (1972, p. 22) selected a transient-period length and checked whether the simulation result corresponded with the known, analytical solution. (Next they changed the model to represent a more realistic system; also see validation in Section II.5.) Fortunately, if the total timepath is long we may suppose that the bias caused by transient observations is negligible. Recently Blomqvist (1970) derived a very interesting result. He assumed a single-server queuing system with independent arrival and service times, a FIFO queuing discipline and an "empty" starting state (i.e., no customers in the system). Then he proved that the mean square error of the estimated mean steady-state waiting time is minimal if <u>all</u> observations (transient and steady state) are averaged. This result depends on two more assumptions. (a) The runlength is "large." (b) The variation coefficient, σ/μ, is at least one. (Condition (b) was proved to hold for a number of simple queuing systems.) A different procedure, eliminating initial observations, is derived by Fishman (1971, p. 30); a completely different approach is follwed by Crane and Iglehart (1972b) as we shall see in a moment.

If we throw away the transient observations then the computer time needed to pass through the transient state is wasted. We may try to minimize this <u>wasted computer time</u> by the choice of adequate starting conditions; cf. Meier et al. (1969, p. 297) for a case study. Yet the empty state is often selected as the initial condition. Starting in the empty state usually prolongs the transient state, but, on the other hand, it simplifies the simulation program. Once the initial conditions are selected Conway (1963, p. 51) and Nelson (1966, p. 15) recommend using the same starting conditions for each simulated system variant in order to obtain fair comparisons; Meier et al. (1969, p. 307-308) discuss deviations from this general rule. A discussion of the initial conditions can also be found in Fetter and Thomson (1965, p. 692), Hillier and Lieberman (1968, pp. 462-463), and Huisman (1969, p. 9).

Crane and Iglehart (1972b) propose a procedure that eliminates the problem of the determination of the transient-period length. As an example consider a single-server queuing system. Start the simulation in the empty state and average all observations, transient and steady state ones, until the system <u>returns</u> to its empty state. Then the next observations are independent of the past history! So blocks of observations are formed, the (stochastic) block size depending on the return to the empty state. (For an exact definition of the block size see Chapter VA.) The important advantage of this approach is that the blocks are independent so that the analysis is simplified; also see Section II.9. At the same time this approach permits recording observations from the very start of the simulation!

(ii) Transient-state simulations. Nelson (1966, p. 14) points out that until now most studies have concentrated on steady-state performance, while actually most problems are of a nonstationary type. Many systems never reach the steady state since they are closed down at the end of a fixed period of time, or they are subjected to all kinds of shocks; e.g., a service bureau that is closed down at the end of each day and a national economy, respectively. More examples of transient behavior can be found in the literature[6]; we especially recommend Emshoff and Sisson (1971, pp. 189-202).

As by definition the initial conditions do influence the system response in the transient state, the role of the starting conditions is completely different in studies of transient behavior as compared to studies of stationary behavior. We can take the "historical" starting conditions like the empty state for a system that is reopened, or the conditions in a recent year for which data are available when a system is subjected to shocks (Orcutt et al., 1961, p. 131). The initial conditions may also be a subject of study themselves, for we may investigate how the system reacts to different starting conditions. With Nelson we expect an increase in the number of simulation studies of transient behavior. We shall return to the starting conditions in Sections II.8 and II.9 where we shall discuss the influence of the initial conditions on the determination of the sample size and the analysis of the output.

Another peculiar aspect of simulation is as follows. The actual experimentation in simulation consists of running the computer program. The typical difference between a simulation and a real-world experiment is that in simulation we can manipulate sampling without introducing bias in the response of interest. This manipulation is done by variance reduction techniques, also called Monte Carlo techniques. These techniques originate from outside simulation (e.g., from sample surveys and the estimation of integrals). In Chapter III six techniques will be discussed that may be applied in complex simulations: (1) stratified sampling: the responses of the (replicated) simulation runs have different weights, the known weights depending on the strata to which the random numbers belong; (2) selective sampling: the input variables are sampled in such a way that their frequencies agree with the theoretical, expected frequencies; (3) control variates or regression sampling: the average values of the input variables are compared with their theoretical means and the response is corrected for deviations between the empirical averages and the theoretical means; (4) importance sampling: the original distributions of the input variables are replaced by new ones (so that those input values resulting in important responses obtain higher probabilities) and the response is corrected for this distortion; (5) antithetic variates: a partner run is generated from the antithetic random numbers $(1 - r)$; the resulting negative correlation between the two runs decreases the variance of the response averaged over runs; (6) common random numbers: two or more system variants are simulated using the same random numbers so the variants are studied "under the same circumstances." This type of variance reduction should be distinguished from variance reduction due to the increase of the sample size or an efficient choice of the combinations of the levels of the factors in the experiment. The latter two techniques will be discussed in Sections II.8 and II.6, respectively.

II.5. VALIDATION

After the above steps we are ready to validate the model. As we explained in Section II.3 we can feed in historical data, have them processed by the computer program, obtain simulation output and compare the simulation output with historical output. For validation it is necessary that at least one variant of the simulated systems exists in real life so actual data are available. We point out that we may simulate an existing system not because we are interested in the simulation output, but only to validate the model. Once this model is validated we change the model in such a way that it hopefully represents the system of interest in a realistic way; see, however, the remarks in Schmidt and Taylor (1970, pp. 499-501). A more philosophical discussion of validation is given by Naylor et al. (1967a, pp. 310-320) and Naylor and Finger (1967); also see Gregg and Simon (1967).

Naylor and Finger (1967) list several possible measures and techniques that can be used in validation. As _measures_ they mention the average value of a simulated timepath, the whole timepath itself, the amplitude, turning points, etc. Both Naylor et al. (1967a, p. 318) and Naylor and Finger (1967) list a number of _techniques_ for testing how well the simulation output agrees with the historical output. They mention the well-known _chi-square_ and the _Kolmogorov-Smirnov tests_ for comparing actual frequencies with theoretical frequences. [Fishman and Kiviat (1967, p. 20) discuss Cochran's variance test as an alternative to the chi-square test.] _Factor analysis_ and _spectral analysis_ can be applied to the simulation output and to the actual output to see if the factor loadings and spectra are the same for the simulation output and the actual output, respectively. (We shall briefly return to spectral analysis in Section II.9.) It is possible to perform simple _regression analysis_ between the actual and the simulation output and to test if the intercept is zero and the slope is 1; Aigner (1972), however, questions this procedure. Theil's "_inequality coefficient_" can be calculated or the variant shown in de Wolff (1967, p. 338-342).[7] _Turing's test_ is applied in the simulation of human problem solving.[8] Mihram (1972a, pp. 250-254, 307-310; 1972b) discusses these and some more measures and techniques, e.g., testing the _average_ simulation output vs actual output applying _parameteric or nonparametric tests_. Howrey (1972, pp. 24-33) mentions the problems of using _loss functions_ for the evaluation of a model; he lists some more measures and amply discusses the turning-point measure (much used in econometric modeling) and the spectral-analysis technique. Hanna (1971) introduces information-theoretic measures for evaluating simulation models. (He concentrates on artificial intelligence simulation; also see Starbuck, 1971.) It will be clear that many validation aspects are not specific for simulation models but also hold for other model types. A specific aspect of simulation is that it gives a complete timepath consisting of autocorrelated data.

We observe that goodness of fit tests like the chi-square test, can also be used to compare actual and theoretical input data. Naylor and Finger (1967, p. 100) point out that the latter testing is useful since otherwise we would construct an unrealistic model (since its input distributions are unrealistic). Without testing the input distributions we would discover that the model is unrealistic in the validation phase. In the meantime we spent time constructing, running, and testing this unrealistic model! Related to validation and testing input data is "verification" discussed by Fishman and Kiviat (1967, pp. 3, 11-19). They describe verification as testing whether the model, especially parts of the model, behaves as we assume it does (while validation means that we test if the model gives the same output as the real world system). We may test if the random number generator gives independent numbers as we assume it does. We can verify the complete model (instead of its parts) by considering a simplified, possibly deterministic, problem for which an analytical solution exists. For this simplified problem can also be simulated and its output compared with the analytical solution; see Emshoff and Sisson (1971, p. 170), Mihram (1972a, pp. 245-250), and Schmidt and Taylor (1970, pp. 502-507). (This simplified model can further be used for a preliminary sensitivity analysis.) Verification or "debugging" of the computer program is possible doing some calculations by hand.

Validation is related to sensitivity testing and evoluationary model building. If the model is sensitive to the exact input specification then additional information may be collected to construct a valid model. If a model is found to be invalid a next cycle in the modeling process may start. Observe that because of the building block approach in most simulation studies, it is sensible to validate submodels for system components. Several good surveys of these aspects can be found in the literature.[9)]

Obviously a model or model part may be erroneously rejected in the validation or verification phase if the sample size is small so large sampling error is possible. Besides this so-called α-error or type I error there is the β-error or type II error, i.e., the probability that a false model or model part is erroneously accepted. Orcutt et al. (1961, p. 32) give a detailed discussion of the deviations between the actual and the simulation output originating from model specification error and sampling error, the latter type of error resulting in α and β errors; also see Stash (1968). Additional references on validation are Howrey (1972) giving an interesting survey of validation of linear econometric models, and Dutton and Starbuck (1971a; see index on p. 706, "prior and posterior uses of data"), etc.[10)] Case studies with explicit attention to validation are given by Gürtler (1969, pp. 106-109), Pegels (1969, p. 228), and Van Horn (1971, pp. 253-256). The former two case studies do not use formal statistical tests for the comparison of model output and historical output.

II.6. EXPERIMENTAL DESIGN

If in the validation step the computer simulation model is not rejected, we may proceed to using the computer program for simulation of the various systems of interest. These systems (or system variants) differ from each other insofar as they have different values or types of system parameters, input variables, and behavioral relationships (also called operating characteristics). These varying parameters, variables, and relationships are called factors in statistical design terminology. In order to find the effect of a factor, this factor must be varied or, in design terminology, considered at several levels. Obviously a quantitative factor could be considered at many levels. An example of a quantitative factor is the parameter, say λ, of the exponential distribution in a queuing model.

A qualitative factor can be considered only at a limited number of levels. An example is the factor "priority rule" in a queuing problem where we may decide to consider two levels, say the "first in, first out" and the "random" queuing discipline. Notice that the numbers of the levels of a qualitative factor are just a convenient shorthand notation and are not used to formulate a mathematical function between the factor levels and the response.

If we want to investigate k factors, factor i $(i = 1,\ldots,k)$ having L_i levels, then the number of combinations of factor levels is $L_1 \times L_2 \times \cdots \times L_k$. The problem is that even for just a few factors this number of combinations is high. For example, if there are only seven factors and each factor is at its minimum number of levels (i.e., $L_i = 2$) then we have $2^7 = 128$ combinations! Therefore we shall try to limit the number of combinations that will actually be investigated. (Remember that this investigation means that the selected factor combination specifies one variant of the system which is simulated over time and results in one timepath for that particular system variant.) This is the area of "experimental design." We find it convenient to distinguish the following parts of experimental design.

(i) Determination of the important factors or "screening." As we have seen above the number of factor combinations rapidly increases with the number of factors considered in an experiment. Therefore we may decide to concentrate on the important factors. A preliminary investigation is necessary to screen for important factors. Especially suited for investigating many factors in relatively few observations are the following designs.

2^{k-p} designs: All k factors are at two levels and only a fraction (a 2^{-p} fraction) is examined. Depending on the selected fraction estimates are still possible of main effects and (low-order) interactions among the factors. If k is high the number of combinations is still too high (i.e., 2^{k-p}) so other screening designs are more suitable.

Random designs: The combinations of factor levels are randomly selected from among all possible combinations. The number of combinations, say N, can be determined independently of the number of factors and levels. So N may be chosen even smaller than k. A disadvantage is that the random selection of combinations may result in combinations that do not permit "good" estimates of the individual effects. (More technically phrased: the degree of orthogonality of the columns in the matrix of independent variables is stochastic instead of being controlled.) Therefore the following type of design was developed.

Supersaturated designs: The number of combinations (N) is smaller than the number of factors (k) and the combinations are so selected that (given N and k) the estimators of the factor effects are as "good" as possible. (The maximum nonorthogonality of the design columns is minimized.) If there are very many factors then the following designs are more suitable.

Group-screening designs: The k factors are combined into g groups of factors $(g \ll k)$ and these g group factors are examined in a 2^{k-p} or supersaturated design; only those group factors that are found to be important are next split into several groups of smaller size (ultimately of size 1).

In Chapter IV these four design types will be discussed in detail and compared with each other.

(ii) Further investigation of the important factors. We have to decide which factors, which levels, and which combinations of levels, will be studied in the experiment. The choice of the factors may be based on the above screening phase if there are many conceivably important factors. The number of levels will be restricted in order to keep the number of combinations small; often we take just two or three levels. If only a few combinations result we may use a full factorial design, i.e., all combinations are investigated. Otherwise incomplete designs can be applied. Two simple types of incomplete designs are as follows:

<u>2^{k-p} designs</u>: These designs were discussed as screening designs. They can also be applied in a more detailed study of the factors since larger fractions permit further study of the k factors. For, a small fraction yields estimators of main effects only (biased if certain interactions exist); larger fractions give estimators of main effects and interactions between no more than two factors (unbiased if no interactions among three or more factors exist).

<u>Response surface designs</u>: These designs combine the 2^{k-p} designs with suitably chosen additional combinations so that the response can be estimated as a function (a first or second degree polynomial) of the independent variables. (All k factors are assumed to be quantitative.)

Both 2^{k-p} and response surface designs can be easily used in a "<u>sequential</u>" way, i.e., we obtain observations for a few combinations of factor levels, then we analyze these observations and only after this analysis we decide for which combinations (old or new ones) additional observations will be generated. These new observations are again analyzed (usually together with the old ones) before we decide which observations will be generated next, etc. In a 2^{k-p} design we can first take one particular small fraction, analyze the observations and if this analysis indicates that this fraction is too small to estimate all important effects, then we can extend the fraction to a bigger one which permits estimation of all these effects. Response surface designs are applied in "<u>response surface methodology</u>" (or RSM), a technique that tries to find the <u>optimum</u> combination of the levels of k quantitative factors in the following steps.

(a) <u>Experimental design</u>, i.e., selection of the combinations of factor levels. RSM uses full and fractional 2^{k-p} designs and special RSM designs.

(b) <u>Fitting a regression equation</u> to the observations. This regression equation is called the "response surface" and expresses the response as a function of the independent variables.

(c) Climbing the response surface towards the top. To find the direction of increasing responses the "steepest ascent" method is usually applied. In the direct where higher responses are expected, steps (a), (b), and (c) are repeated until the top is apparently reached.

(d) Canonical analysis of the apparent maximum response. This analysis reveals if we have a unique maximum, several maxima, a saddlepoint or a ridge.

In Chapter IV a detailed description will be given of 2^{k-p} designs, their sequentialization and their analysis. For RSM we refer to the bibliography in Chapter IV and the surveys in Hill and Hunter (1966) and Herzberg and Cox (1969). We emphasize that experimental design as described above is just a part of the design of a computer simulation experiment, which further comprises specification of starting conditions, variance reduction techniques, sample size determination, etc.

II.7. ANALYSIS OF SEVERAL SYSTEM VARIANTS

The analysis and the design of an experiment can be distinguished but usually not separated. As we observed in Section II.6 we prefer a sequential design and analysis of simulation experiments. A pure sequential approach would reanalyze the output after each additional observation. In practice, however, the analysis is usually performed after several observations have been obtained; this is called a multistage approach. Sequential designs have become more important. For the origin of most experimental designs is in the agriculture where sequentialization is difficult since one has usually to wait a whole year before the next planned observations come available. In more recent years experimental designs have also been used in the industry where shorter periods of time are needed for the generation of observations. Sequential designs have been developed and applied especially in the chemical industry. Computer simulation experiments are by their very nature well-suited to sequential designs since the computer operates in a sequential way!

There are many sequential designs; references are given in Naylor et al. (1967a, p. 338), Naylor et al. (1967b, pp. 1329-1330), and Wetherill (1966).

After having pointed out the interrelations between design and analysis we shall emphasize the analysis in the rest of this section. We shall discuss some techniques that can be applied when several system variants are compared with each other. Suppose we want to study k system variants, or briefly k systems. Each system corresponds with one particular combination of the levels of the factors varied in the experiment. Each combination specifies one population from which observations are sampled. We shall discuss situations where the number of observations per population is not fixed but is determined during the experiment, and situations with a fixed number of observations.

(i) Multiple ranking procedures. Procedures exist for determining how many observations should be taken from each of the k (≥ 2) populations in order to select the best population. The best population is usually the one with the largest mean or, depending on the problem, the smallest mean. Some procedures have been derived for other selection criteria than the mean, e.g., the variance, or for other problem formulations than selection of the best population, e.g., a complete ranking of all populations from worst to best. Procedures for these complete or incomplete rankings are called multiple-ranking procedures; in the literature they are also found under such headings as multiple selection and decision procedures. Most procedures are based on the "indifference zone" approach, i.e., large samples would be required to select the best population if the population means differ only slightly, while the loss involved in a wrong selection would then usually be small; therefore the procedure guarantees a correct selection with probability at least P^* (or $1 - \alpha$) only if the best population mean is at least δ^* better than the next best mean (P^* and δ^* are specified by the experimenter.) Most ranking procedures are sequential. Chapter VC, contains a survey of existing procedures, a study of their efficiency and robustness, and some heuristic procedures. (The robustness of a procedure is its sensitivity to violations of the underlying assumptions like normality, independence, etc.)

(ii) Response surface methodology. If all factors are quantitative then RSM can be applied to find the best system; see Section II.6 above.

(iii) Multiple comparison procedures. The number of observations per population may be fixed, because we are doing a pilot experiment or because the available computer time is limited. We can make various types of comparisons among the population means, e.g., comparisons with the "standard" mean that corresponds with the existing system (or $\mu_i - \mu_0$ with $i = 1, \ldots, k-1$); all pairwise comparisons ($\mu_i - \mu_{i'}$ with $i \neq i'$); comparisons of the average response of $(k - 1)$ experimental systems with the standard system (or, more general, linear contrasts $\Sigma_{i=1}^{k} c_i\mu_i$ with $\Sigma_1^k c_i = 0$); study of the means μ_i themselves; selection of a subset containing the best population or, in other situations, a subset containing all populations better than the standard population. (The populations in the subset may be further studied in additional experiments.) In multiple comparison procedures the probability is at least $(1 - \alpha)$ that all statements based on a single experiment are correct. For example, all $k(k - 1)/2$ confidence intervals for pairwise comparisons are correct with probability at least $1 - \alpha$ (α being the so-called experimentwise error rate). Notice that if all factors are quantitative then a regression analysis is more efficient than a multiple comparison procedure. We refer to Chapter VB, for a discussion of multiple comparison procedures for various situations, their efficiency and robustness, and types of error rates.

II.8. SAMPLE SIZE

In Section II.7 we have already discussed the determination of the sample size, i.e., the number of observations per population. In this section, and the next one, we shall investigate some aspects of the sample size that are specific for simulation. But first we shall make a few general statements. The sample size effects the "statistical reliability" of the estimated response of the simulated system. This reliability may be measured by the standard deviation of the mean response for the system variant being simulated. The reliability can be increased by taking a larger sample. Suppose we are interested in the mean response μ and we have n independent observations of that mean response, say $\underline{x}_1, \underline{x}_2, \ldots, \underline{x}_n$. It is well known that the standard deviation of $\bar{\underline{x}}$, the average of the n independent observations, is $\sigma/\sqrt{n}$ where σ is the standard deviation of an individual observation $\underline{x}_i$ $(i - 1, \ldots, n)$. Hence in order to, say, half the standard deviation of this average, we have to take a sample size four times as big as the original one. This shows that large increases in the sample size are needed to reduce the stochastic fluctuations. (As we mentioned in Section II.4 variance reduction techniques can be applied in order to mitigate the need for large sample sizes.) The above reasoning needs adjustment in simulations with dependent observations as we shall see later.

There are several possibilities for increasing the sample size in a simulation experiment. These possibilities vary with the following two dichotomies, introduced by Gafarian and Anker (1966).

(i) Time-slicing vs event-sequencing programming. Time-slicing means that we divide the period of time to be simulated, into small slices; at the end of each time slice we determine if and how the system changed. If the system changes continuously this is the only programming technique possible. Many systems, however, change their state only at certain instants of time. Then we can use either the time-slicing or the event-sequencing technique. In event-sequencing we jump from one instant of time to the next instant of time where the system changes its state. For instance, in a simple queuing problem the next instant of time where something happens to the simulated system, is the instant where either a customer arrives into the system or the servicing of a customer ends; also see the example in Naylor et al. (1967a, pp. 126-136). A detailed discussion is given by Emshoff and Sisson (1971, pp. 159-169) and

Mihram (1972a, pp. 228-231). Note that Nance (1971) refines the distinction between these two programming approaches and compares the efficiency of the resulting (four) procedures. In Section I.6.2 we observed that Forrester's industrial dynamic approach is also a time-slicing simulation, but concentrated on highly aggregated and deterministic models; see Meier et al. (1969, pp. 80-117), Niedereichholz (1971), and Smith (1970, pp. 769-770).

(ii) Terminating vs nonterminating systems. In a terminating system the simulation run ends if a specified event occurs. Gafarian and Anker (1966, p. 27) give as examples a duel and an equipment failure model where simulated time ends when one or both contestants are killed, and when the equipment breaks down, respectively. In a nonterminating system there is no critical event that stops the simulation run; compare a telephone exchange. We refer to Gafarian and Anker (1966) for a further discussion of the relations between fixed and variable time increment programming, terminating and nonterminating, continuous and discrete systems. The rest of this section, however, deviates from their article. Observe that one simulation "run" is one movement of the system from the beginning to the end of the simulated period of time.

There are systems that are physically nonterminating but that behave exactly as terminating systems as far as sample size considerations are concerned. We have three important types of such systems in mind. First, there are systems that close down at regular points of time. For example, at the end of the day a bank is closed, the last customers are served and the next day the bank opens again. Then one run comprises one day. Second, consider systems for which the response after a fixed period of time is to be estimated for various strategies (these strategies giving shocks to the system). An example is a firm for which we want to find the profit in the coming three months for various investment alternatives. So the "specified event" that terminates the simulation run is the arrival at the end of the planning period. The third type comprises physically nonterminating systems for which we want to study transient behavior (e.g., as a function of the initial conditions). As soon as the system reaches the steady state the simulation is terminated. Since we classify these three types of physically nonterminating systems as terminating systems, it will be clear why from now on we shall assume that in nonterminating systems we are interested only in steady-state characteristics; also cf. Section II.4 above.

We shall discuss three ways for increasing the sample size [Meier et al. (1969, p. 300) distinguish more ways but the basic differences are covered by our classification]: (i) Smaller time-slice; (ii) Replication of simulation runs; (iii) Continuation of the simulation run. Obviously the first alternative can be applied only if time-slice programming is used. A smaller time slice implies that the system is observed more frequently in a given period of time. Gafarian and Anker (1966, p. 41) derived that in most situations this way of increasing the sample size is less efficient. We would add that besides being inefficient manipulation of the time-slice is often impractical. For as Fishman and Kiviat (1967, pp. 25-26) show, the length of the time slice may be more or less fixed by the type of system activities; also see Meier et al. (1969, p. 300). Further time-slicing itself is an unattractive programming technique if the system is easier programmed using event-sequencing or if it takes less computer time to run the system by event-sequencing. An example is a simple queuing problem; a counter-example is a freeway diamond interchange which is difficult to program in event-sequencing since it is hard to define the events. Finally, time-slicing is impossible if we want to simulate in an available simulation language that is based on event-sequencing. So from now on we shall ignore the length of the time slice as a means of sample size increase.

For terminating systems replication of runs is the single remaining means, since the length of the run cannot be manipulated. If we use a new set of random numbers for each run then each run yields an independent observation so traditional statistical techniques can be used to analyze the simulation experiment. (Replicated runs also simplify the application of variance reduction techniques; see Chapter III.) We point out that Bruzelius (1972a) failed to make the distinction between terminating and nonterminating systems; he simulated a bank during $83\frac{1}{3}$ hours without interruption. In Chapter VA, we shall further discuss appropriate analysis techniques like sample size determination for confidence intervals of predetermined lengths or tests with predetermined α and β-errors, for one mean or for the difference between two means.

For nonterminating systems both replicated runs and continued runs are possible. Gafarian and Anker (1966, p. 41) concluded that replication of runs is more efficient, assuming positive serial correlation among the individual observations (a reasonable assumption for many simulations). If we assumed that the system were in its steady state at the very beginning of the simulation then it would be easy to show that replication indeed reduces the variance of the average response; see Exercises 1 and 2. Actually the only case study we could find in the literature where the system is in its steady state at the beginning of the simulation is the "closed-loop feedback control system" discussed by Hauser et al. (1966, p. 79). In general, however, the simulated system has to pass through the transient state before it reaches the steady state. Consequently if we are interested only in the steady state we usually throw away the transient-state observations. (However, as we saw in Section II.4 the Crane-Iglehart procedure uses all observations.) If we replicate runs we throw away transient observations at the beginning of each run. This problem is also discussed by Conway (1963, p. 55), Conway et al. (1959, pp. 109-110) and Tocher (1963, pp. 176-177). Unfortunately, at present no solution is available for choosing between replicated and continued runs, weighing the variance reduction against the computer time wasted in the transient state (or the transient-state bias). Moreover, Mihram (1972a, pp. 448-450) points out that estimating the moments from a single continued run may give incorrect estimates of the population moments, namely in case the simulation-generated time series is not "ergodic."[11) There is one more factor relevant for the choice between continued and replicated runs, viz., the analysis of the simulation output. The analysis of replicated runs is very simple since each run gives one independent observation. A continued run gives observations that are serially correlated and this complicates the analysis considerably. This analysis problem will be discussed in the next section. Before proceeding to that section, we summarize the arguments in favor of replication of runs (also see Mihram (1972a, p. 449)):

(a) Many systems are terminating and consequently do not enable continuation of the simulation run.

(b) Independent runs permit classical analysis techniques.

(c) Ergodicity problems are eliminated.

(d) Replication gives lower variance assuming no (or not much) initialization effort.

II.9 ANALYSIS OF STEADY-STATE OUTPUT AND RUNLENGTH

In this section we shall discuss the analysis of the output of a nonterminating system for which we want to estimate the expected steady-state response. We shall assume that--in case we do not apply the Crane-Iglehart procedure--we are able to discard the transient-state observations so the remaining timepath consists of steady-state observations only.[12] There remains the problem of the analysis of this steady-state output. We shall discuss replicated and prolonged runs separately.

(i) Replicated runs. We may decide to replicate runs despite computer time possibly being wasted in the transient state. Each run gives one independent estimated average $\bar{x}$. The "stationary r-dependent central limit theorem," formulated in Mechanic and McKay (1966, p. 4) and in Chapter VA, implies that the average of each run is (approximately) normally distributed (unless periodicities exist). Hence the Student t-statistic can be used to construct a confidence interval of the form $(\bar{\bar{x}} \pm t_\alpha s_{\bar{x}} / \sqrt{n})$ where $\bar{\bar{x}}$ is the overall average of the averages $\bar{x}_i$ and $s_{\bar{x}}$ is the standard deviation calculated from the $\bar{x}_i$ $(i = 1, \ldots, n)$.

(ii) Continued run. There are three approaches to the analysis of a prolonged run.

Approach 1: "Independent" subruns. Divide the long, continued run into, say, n subruns (after having discarded the transient observations of the initial phase of the run). Denote the averages of these subruns by $\bar{x}_1, \ldots, \bar{x}_n$. Assume that the serial correlation decreases as the "lag," i.e., the distance between the individual observations x, increases. (This assumption implies that there are no periodicities.) Consequently, only the first "few" observations of subrun i will be correlated with the last observations in subrun i - 1 and therefore the averages $\bar{x}_i$ and $\bar{x}_{i-1}$ will show only small correlations. If the subruns are long enough the correlations among subrun averages can be neglected for practical purposes. Confidence intervals based on the t-statistic are similar to the ones discussed for replicated runs. Mechanic and McKay (1966) derived a procedure for determining the length of the subruns in such a way that the subrun averages are approximately independent. Their procedure is iterative; it estimates the correlation between the subrun averages and takes larger subruns until this correlation is "small" enough. The smallness of the correlation coefficient is judged against an experimentally derived criterion, given in Mechanic and McKay (1966, p. 20). They applied their rule to several queuing problems and derived some theoretical relations, both with satisfying results for the procedure.

Approach 2: Estimation of serial correlations. This alternative does not create subruns but estimates the variance of $\bar{x}$, the average based on the whole run. In the expression for this variance serial correlation coefficients occur.[13) So the correlations among the individual observations must be estimated. Different estimation formulas are used by Fishman (1967) and Hauser et al. (1966).

Approach 3: Independent blocks. In Section II.4 we briefly described the approach followed by Crane and Iglehart (1972b).

Creating independent blocks (or "tours") has been proposed by Kabak (1968). The analysis of these blocks has been investigated by Crane and Iglehart (1972a, 1972b) and Fishman (1972).

In Chapter VA a more detailed account will be given of these and other techniques for the estimation of the reliability of the estimated mean steady-state response. The estimated variance can be used to determine (sequentially) whether the (total) run is long enough to satisfy the required reliability of the estimated simulation response, analogous to the analysis of terminating systems.

It may be that besides the mean response and the variance, we want to know the dynamic characteristics of the steady-state response (e.g., periodicities). For as Kiviat (1967, p. 7) and Mihram (1972a, pp. 261-262) point out, we may distinguish: (i) Static measures or measures of average performance, like means, standard deviations, and histograms; (ii) Dynamic measures like the autocorrelation or the spectrum. Autocorrelation coefficients were studied by Hoggatt and Holtbrugge (1966) in his market simulation. As Fishman and Kiviat (1967, pp. 15-16) remark the spectrum is just a (Fourier) transformation of the autocorrelation so it provides the same information but they find it convenient to work with the spectrum. For further discusstion of spectral analysis and additional literature we refer to Anderson (1971), Fishman (1967, 1968), Fishman and Kiviat (1967), Mihram (1972a, pp. 175-179, 467-483), and Naylor (1971, pp. 35, 247-268, 312-317).

Because of the emphasis on spectral analysis in the publications of Fishman and Kiviat, two authoritative authors on simulations, it seems useful to indicate the limitations we see for this technique.[14)] As it follows from Fishman (1968, p. 17) spectral analysis is applied to steady-state observations. If the simulation output satisfies these conditions then spectral analysis is a most useful technique (if at least we want to perform a more sophisticated analysis where both static and dynamic properties of the system are studied). Note that a time series (possibly generated by a simulation experiment) may be made stationary by an appropriate transformation (e.g., taking increments or moving averages) that removes a possible trend (shown by the mean or variance); see Mihram (1972a, pp. 445-459) and for a detailed study see Anderson (1971). However, in Section II.4 we pointed out that we expect an increase in the number of simulation studies of transient behavior. So we expect that traditional analysis techniques based on independent observations, will remain useful for the analysis of the output of a simulated system. Moreover, these techniques are necessary for the design and analysis of the simulation of several systems. For in a simulation study we shall be interested in several variants of a system. Deciding which variants we shall study implies experimental design as discussed in Section II.6. The analysis of the output of several systems can be done applying analysis of variance, multiple comparisons, multiple ranking, response surface methodology, etc. All these techniques assume independent observations.

II.10. MISCELLANEOUS STATISTICAL ASPECTS

There remain more statistical aspects of simulation. For example, the design and analysis of an experiment is complicated if we have more than one output variable of interest. Naylor et al. (1967a, pp. 339-440) observe that multiple responses form an underdeveloped area of statistical design and analysis and that practitioners solve this problem by considering the experiment with multiple responses as multiple experiments with a single response. (In Chapter VI we shall present a case study where we also have multiple responses, as a multiple ranking procedure is studied for several reliability levels P^* each level giving a response.) Theoretically several systems may be studied in a simulation in order to select the optimal system, the selection being based on a criterion. This criterion is a function which combines the multiple responses into a single value so that a choice can be made, at the same time eliminating the problem of multiple responses. Naylor (1971, p. 28) mentions Fromm's utility function for comparing several economic policies. In practice one might not be willing or able to specify a criterion function. Further, in preliminary studies where we just want to explore possible relationships, a criterion function is usually also missing. For a discussion of the problem of multiple criteria we also refer to Bakes et al. (1970, pp. 47-50). We may indeed be confronted with multiple responses. We shall briefly return to the multiple response problem in the following chapters.

There are more statistical aspects of simulation than we have already discussed. For example, Conway (1963, pp. 59-60) gives several rules for deciding, after the sample size N has been determined, which individual "temporary entities" in the system, like customers, should be observed (e.g., the first N customers that arrived after some point of time or the first N customers that left after that point of time). Conway (1963, p. 54) also discusses how to measure the performance of the "permanent entities," like the service counters. Orcutt et al. (1961, p. 377) mention the possibility of forecasting the system response not directly from the simulation. Instead a regression equation can be used to forecast the output, with as explanatory variables only the simulation output or both simulation output and some other data, e.g., aggregated economic variables measured by the national government. Nelson (1966, pp. 19-20) observes that indeed we should plan how the data will be analyzed before the experiment is performed. However, often it is not known beforehand which data will be needed. It is better to have too much data than too little. Therefore the computer can record the output on tape, and this tape can be used later on as input for an output-analysis program.

Finally, we point out that the statistical reliability should be distinguished from the numerical accuracy. Lombaers (1968, p. 250) mentions the following factors for the accuracy.

(i) The number of significant digits used in the calculations. With Hauser et al. (1966, p. 84) we observe that the loss of significant digits can be reduced when calculating the variance by considering $(\underline{x}_i - a)$ instead of $\underline{x}_i$ (where a is a suitably chosen constant), since the variance of $\underline{x}_i$ is equal to that of $(\underline{x}_i - a)$.

(ii) Interpolation in tables. As Naylor et al. (1967a, p. 250) remark some simulation languages, e.g., GPSS, use interpolation in tables when sampling stochastic variables. Such interpolation is also applied by Orcutt et al. (1961, pp. 123-124).

II.11. LITERATURE

We hope that this chapter will help the reader interested in a particular statistical aspect of simulation, to find the appropriate literature. If he has a general interest in the statistical aspects of simulation then we recommend Naylor et al. (1967a) and Naylor et al. (1967b). These two publications give an introduction to the various facets of the design and analysis of computer simulation experiments and they contain many more references. For a very extensive list of literature we refer to the category "formal and statistical methods" in the bibliography drafted by Dutton and Starbuck (1971b, pp. 164-167); also see Dutton and Starbuck (1971a, pp. 589-594, 693-699, 705-706), and Meier et al. (1969, pp. 299-330). We hope that our own study will also help to explain the statistical implications of simulation. For in this study we have tried to cover the many statistical aspects of simulation in some detail. The remaining chapters contain many more references. These chapters concentrate on particular statistical aspects. They give both a survey of existing techniques and some new results. Chapter III covers variance reduction techniques; Chapter IV discusses experimental designs; Chapter V shows the effect of the sample size on the reliability (including multiple comparison and ranking procedures); Chapter VI gives a case study where the robustness of a ranking procedure is studied applying several techniques of the preceding chapters.

EXERCISES[15)]

1. Assume that the individual observations in a simulation run are positively correlated; there is no transient phase. Prove that it is more efficient to generate two independent runs of n observations each, then one prolonged run of $2n$ observations.

2. How could you further improve the efficiency of the above procedure generating two runs instead of one prolonged run? (Hint: see Section II.4.)

3. One wants to simulate two system variants with positive correlation. How can this be done? What is the advantage? How can the difference between the average system responses be tested?

4. Four systems are simulated, giving estimated means $\underline{\hat{\mu}}_i$ $(i = 1, \ldots, 4)$. Student's t-statistic is used to give confidence intervals for $\mu_i - \mu_{i'}$ $(i \neq i')$. What is the probability that the confidence intervals for both $\mu_1 - \mu_2$ and $\mu_3 - \mu_4$ hold? Does this joint confidence coefficient also hold for $\mu_1 - \mu_2$ and $\mu_1 - \mu_3$?

5. An inventory system is simulated, the input factor "demand" being sampled from a distribution with mean η. The average demand sampled in a run will be $\underline{\bar{y}}$. The mean output is therefore not estimated by $\underline{\hat{\mu}}$ but by $\underline{\hat{\mu}} - a(\underline{\bar{y}} - \eta)$, so-called control variate technique. Derive the optimal value of the constant a.

6. Naylor et al. (1967b, p. 705) used a steady-state formula to estimate expected total profit during the whole planning period (i.e., inclusive of profit realized in the transient phase at the beginning of the planning period). It is correct to use this steady-state formula?

7. Simulate a simple queuing system. Estimate mean steady-state waiting time, once using all observations à la Blomqvist, once eliminating transient observations applying Conway's rule of thumb; see Section II.4.

8. Test if the simulation model for the simple queuing system gives significantly different mean waiting times when replacing the exponential service distribution by a Gaussian distribution with the same mean.

9. Select a model with a known analytical solution, e.g., a simple queuing or inventory model, or an econometric difference equation. Simulate the system and verify the simulation model.

10. Make it plausible that antithetic variates applied to a single-server FIFO queuing system create negative correlation between the average waiting times of the two runs, one run using r_1, r_2, ..., the other run using $(1 - r_1)$, $(1 - r_2)$, ...

NOTES

1. A summary of this chapter was published in Management Informatics; see Kleijnen (1972).

2. See Chambers and Mullick (1968, p. 100), Dutton and Starbuck (1971a, p. 705), Emshoff and Sisson (1971, p. 49-59), Kiviat (1967, p. 19-22), Mihram (1972a, pp. 209-260), Naylor et al. (1967a, pp. 23-41).

3. See Chambers and Mullick (1968, p. 106), Emshoff and Sisson (1971, p. 228), Harling (1958, p. 17), Hillier and Lieberman (1968, p. 444), Kiviat (1967, p. 31), Mihram (1972a, p. 236) and Naylor et al. (1967a, p. 69).

4. Emshoff and Sisson (1971, p. 190) give a different definition of steady state (because of possible cyclical behavior).

5. Blomqvist (1970) proved for simple queuing systems that the waiting times of the transient phase have monotonically increasing means and standard deviations.

6. See Chambers and Mullick (1968, p. 101), Emshoff and Sisson (1971, pp. 190, 192, 239), Marks (1964, p. 256), Mihram (1970, p. 89), Saleeb and Hartley (1968), Schmidt and Taylor (1970, pp. 346, 436), Smith and Shah (1964, p. 219) and Van Horn (1971, p. 252).

7. Theil's coefficient has the form

$$\left\{ \sum_{i=1}^{m} \sum_{t=1}^{n} (\underline{y}_{it} - \tilde{\underline{y}}_{it})^2/mn \right\}^{1/2} \bigg/ \left\{ \Sigma\, \Sigma\, \underline{y}_{it}^2/mn \right\}^{1/2}$$

where $\underline{y}_{it}$ and $\tilde{\underline{y}}_{it}$ denote the actual and the predicted variable i in period t, respectively; see Driehuis (1972, p. 185).

8. A "third party" has to find out whether the answers are given by a human being or a computer; if he cannot distinguish between the two types of answers then the simulation is validated; see Mitroff (1969, p. 636).

9. See Elton and Rosenhead (1971, pp. 122-124, 128-219), Emshoff and Sisson (1971, pp. 204-206), Schmidt and Taylor (1970, pp. 498-507), Van Horn (1971). Additional references: Clarkson (1968), Hanna (1971), Lovell (1969, p. 10), Mihram (1970, p. 83), and Starbuck (1971, pp. 678-679).

10. See Bruzelius (1972b), Fagerstedt and Petterson (1972), Fain et al. (1970), Gafarian and Walsh (1966), Kiviat (1967, pp. 48-49), Maisel and Gnugnoli (1972), McKenney (1967), Meier et al. (1969, pp. 294-296), Mihram (1972b), Naylor (1971, p. 119, 153-164), Nelson (1966, p. 19), Schrank and Holt (1967), and Smith (1970).

11. Example (taken from Mihram (1972a, p. 448)): Let the population have characteristics: $\underline{y}(0) = +1$, 0, or -1 with equal probability (1/3), and $\underline{y}(t) = \underline{y}(0)$ for $t = 1, 2, \ldots,$. Then a single run yields estimates $\hat{\underline{y}}_t = +1$, 0, or -1 [depending on the sampled initial value $y(0)$] and $\text{var}(\underline{y}_t) = 0$. For a discussion of ergodicity we refer to Parzen (1962, pp. 72-76).

12. Following Blomqvist (1970) we might decide to use all observations, transient and steady state, assuming that this procedure yields a more efficient estimator. Restricting the analysis to the steady-state observations would then yield a lower bound for the reliability of the estimator. Alternatively we might treat the transient observations as if they were steady state, assuming that they do not affect the analysis significantly.

13. $\operatorname{var}(\underline{\bar{x}}) = \frac{\sigma^2}{N} \left\{ 1 + 2 \sum_{s=1}^{N} (1 - \frac{s}{N}) \rho_s \right\}$ with $\underline{\bar{x}} = \sum_{i=1}^{N} \underline{x}_i/N$,

    $\operatorname{var}(\underline{x}_i) = \sigma^2$ and ρ_s being the serial correlation coefficient with lag s.

14. Compare Dear (1961, pp. 18, 20), Emshoff and Sisson (1971, p. 202), and Van Horn (1971, p. 252).

15. More exercises can be found in Schmidt and Taylor (1970, pp. 507-515).

REFERENCES

Aigner, D.J. (1972). "A note on verification of computer simulation models," Management Sci. 18, 615-619.

Anderson, T.W. (1971). The Statistical Analysis of Time Series, Wiley, New York.

Bakes, M.D., M.J. Bramson, S. Freckleton, P.C. Roberts, and D. Ryan (1970). "Stochastic-network reduction and sensitivity techniques in a cost effectiveness study of a military communication system, Operational Res. Quart. 21 (Spec. Conf. Iss.), 45-67.

Blomqvist, N. (1970). "On the transient behaviour of the GI/G/I waiting-times," Skand. Aktuar., 118-129.

Bruzelius, L.H. (1972a). Estimating Endogenous Parameters in a Dynamic Stimulation Model, in Working Papers, Vol. 1, Symposium Computer simulation versus analytical solutions for business and economic models, Graduate School of Business Administration, Gothenburg (Sweden).

Bruzelius, L.H. (1972b). A Proposal for a General Method for Validating a Simulation Model, in Working Papers, Vol. 1, Symposium Computer simulation versus analytical solutions for business and economic models, Graduate School of Business Administration, Gothenburg (Sweden).

Carter, G. and E. Ignall (1970). A Simulation Model of Fire Department Operations: Design and Preliminary Results. R-632-NYC, The New York City Rand Institute, New York.

Chambers, J.C. and S.K. Mullick (1968). "Strategic new product planning models for dynamic situations," IEEE Trans. Eng. Management, 15, 100-108.

Chan, M.M.W. (1971). "System simulation and maximum entropy," Operations Res., 19, 1751-1753.

Clarkson, G.P.E. (1968). "Letter to the editor," Management Sci., 14, 548-550.

Clough, D.J., J.B. Levine, G. Mowbray, and J.R. Walter (1965). "A simulation model for subsidy policy determination in the Canadian uranium mining industry," Can. Operational Res. Soc., 3, 115-128.

Cohen, K.J. (1960). Computer Models of the Shoe, Leather, Hide Sequence, Prentice-Hall, Englewood Cliffs, N.J.

Conway, R.W. (1963). "Some tactical problems in digital simulation," Management Sci. 10, 47-61.

Conway, R.W., B.M. Johnson, and W.L. Maxwell (1959). "Some problems of digital systems simulation," Management Sci., 6, 92-110.

Crane, M.A. and D.L. Iglehart (1972a). A New Approach to Simulating Stable Stochastic Systems: I--General Multi-Server Queues, Technical Report No. 86-1, Control Analysis Corp., Palo Alto, Calif.

Crane, M.A. and D.L. Iglehart (1972b). Simulating Stable Stochastic Systems: II--Markov Chains, Technical Report No. 86-3, Control Analysis Corp., Palo Alto, Calif.

Dear, R.E. (1961). Multivariate Analyses of Variance and Covariance for Simulation Studies Involving Normal Time Series, Field note 5644, System Development Corp., Santa Monica, Calif.

De Wolff, P. (1967). "Macro-economic forecasting," in Forecasting on a Scientific Basis (Proceedings international summer institute 1966, sponsored by the NATO Science Committee and the Gulbenkian Foundation), Centro de economia e financas, Lisbon.

Driehuis, W. (1972). Fluctuations and Growth in a Near Full Employment Economy, Universitaire Pers. Rotterdam.

Dutton, J.M. and W.H. Starbuck (1971a). Computer Simulation of Human Behavior, Wiley, New York.

Dutton, J.M. and W.H. Starbuck (1971b). "Computer simulation models of human behavior: a history of an intellectual technology," IEEE Trans. Systems, Man, Cybernetics, SMC-1, 128-171.

Elton, M. and J. Rosenhead (1971). "Micro-simulation of markets," Operational Res. Quart., 22, 117-144.

Emshoff, J.R. and R.L. Sisson (1971). Design and Use of Computer Simulation Models, Macmillan, New York.

Fagerstedt, L., and S. Petterson (1972). Validation of Simulation Models--Some Methodological Views, in Working Papers, Vol. 1, Symposium Computer simulation versus analytical solutions for business and economic models; Graduate School of Business Administration, Gothenburg (Sweden).

Fain, W.W., J.B. Fain, L. Feldman, and S. Simon (1970). Validation of Combat Models Against Historical Data, Professional paper No. 27, Center for Naval Analyses, Arlington, Va. (obtainable at Clearinghouse, Springfield, Va.).

Fetter, R.B. and J.D. Thompson (1965). "The simulation of hospital systems," Operations Res., 13, 689-711.

Fishman, G.S. (1967). Digitial Computer Simulation: The Allocation of Computer Time in Comparing Simulation Experiments. RM-5288-1-PR, The Rand Corp., Santa Monica, Calif. (Also published in Operations Res., 16, 280-295, 1968.)

Fishman, G.S. (1967). Digital Computer Simulation: The Allocation The Rand Corp., Santa Monica, Calif.

Fishman, G.S. (1971). "Estimating sample size in computing simulation experiments," Management Sci., 18, 21-38.

Fishman, G.S. (1972). Estimation in Multiserver Queuing Simulations, Technical Report 58, Department of Administrative Sciences, Yale University, New Haven, Conn.

Fishman G.S. and P.J. Kiviat (1967). Digital Computer Simulation: Statistical Considerations, RM-5387-PR, The Rand Corp., Santa Monica, Calif., Nov. 1967. (Published as: The statistics of discrete event simulation," Sci. Simulation, 10, 185-195, 1968. Reprinted in J.M. Dutton and W.H. Starbuck, Computer Simulation of Human Behavior, Wiley, New York, 1971.)

Gafarian, A.V. and C.J. Anker (1966). "Mean value estimation from digital computer simulation," Operations Res., 14, 25-44.

Gafarian, A.V. and J.E. Walsh (1966). "Statistical approach for validating simulation models by comparison with operational systems," 4th Int. Conf. Operational Res. (S.B. Hertz and J. Melese, eds.) Wiley-Interscience, New York.

Gregg, L.W. and H.A. Simon (1967). "Process models and stochastic theories of simple concept formation," J. Math. Psychol., 4, 246-276. (Reprinted in J.M. Dutton and W.H. Starbuck, Computer Simulation of Human Behavior, Wiley, New York, 1971.)

Gürtler, H. (1969). Quantitative Modelle zur Optimierung des Schalterverkehrs in einem Postamt. (Quantitative models for optimizing traffic at counters in a post office.) Doctoral dissertation, Wilhelms-Universität, Münster, Germany.

Hanna, J.F. (1971). "Information-theoretic techniques for evaluating simulation models," in: J.M. Dutton and W.H. Starbuck, eds., Computer Simulation of Human Behavior, Wiley, New York.

Harling, J. (1958). "Simulation techniques in operations research," Operational Res. Quart., 9, 9-21.

Hauser, N., N.N. Barish, and S. Ehrenfeld (1966). "Design problems in a process control simulation," J. Ind. Eng., 17, 79-86.

Herzberg, A.M. and D.R. Cox (1969). "Recent work on the design of experiments: a bibliography and a review," J. Roy. Stat. Soc. Ser. A, 132, Part I, 29-67.

Hill, W.J. and W.G. Hunter (1966). "A review of response surface methodology: a literature survey," Technometrics, 8, 571-590.

Hillier, F.S. and G.J. Lieberman (1968). Introduction to Operations Research, Holden-Day, San Francisco, Calif., Chapter 14.

Hoggatt, A.C., and B.J. Holtbrugge (1966). "Statistical techniques for the computer analysis of simulation models," in Studies in a Simulated Market (L.E. Preston and N.R. Collins, eds.), Institute of Business and Economic Research, University of California, Berkeley, Calif.

Howrey, E.P. (1972). Selection and Evaluation of Econometric Models, in Working Papers, Vol. 4, Symposium Computer simulation versus analytical solutions for business and economic models, Graduate School of Business Administration, Gothenburg (Sweden).

Huisman, F. (1969). Statistische Aspekten van Simulatie (Statistical aspects of simulation), Report no. 2, Afdeling Werktuigbouwkunde, Technische Hogeschool Twente, Enschede (Netherlands).

Johnston, J. (1963). Econometric Methods, McGraw-Hill, New York.

Kabak, I.W. (1968). "Stopping rules for queuing simulations," Operations Res., 16, 431-437.

Kiviat, P.J. (1967). Digital Computer Simulation: Modeling Concepts, RM-5378-PR, The Rand Corp., Santa Monica, Calif.

Kleijnen, J.P.C. (1972). "The statistical design and analysis of digital simulation: a survey," Management Informatics, 1, 57-66.

Kotler, P. (1970). "A guide to gathering expert estimates," Business Horizons, 13, 79-87.

Lombaers, H.J.M. (1968). "Enige statistische aspecten van simulatie" (Some statistical aspects of simulation), Statistica Neerl., 22, 249-255.

Lovell, C.C. (1969). Simulation in Field Testing, P-4152, The Rand Corp., Santa Monica, Calif.

McKenney, J.L. (1967). "Critique of: 'verification of computer simulation models'," Management Sci., 14, 102-103.

Maisel, H. and G. Gnugnoli (1972). Simulation of Discrete Stochastic Systems, Science Research Assoc., Palo Alto, Calif.

Marks, B.L. (1964). "Digital simulations of runway utilization," Operational Res. Quart., 15, 249-259.

Mechanic, H. and W. McKay (1966). Confidence Intervals for Averages of Dependent Data in Simulations II, Technical Report 17-202, IBM Advanced Systems Development Division, Yorktown Heights, N.Y.

Meier, R.C., W.T. Newell, and H.L. Pazer (1969). Simulation in Business and Economics, Prentice-Hall, Englewood Cliffs, N.J.

Mihram, G.A. (1970). "A cost-effectiveness study for strategic airlift," Transportation Sci., 4, 79-96.

Mihram, G.A. (1972a). Simulation: Statistical Foundations and Methodology, Academic, New York.

Mihram, G.A. (1972b). "Some practical aspects of the verification and validation of simulation models," Operational Res. Quart., 23, 17-29.

Mitroff, I.I. (1969). "Fundamental issues in the simulation of human behavior: A case in the strategy of behavioral science," Management Sci., Application Ser., 15, 635-649.

Nance, R.E. (1971). "On time flow mechanisms for discrete system simulation," Management Sci., 18, 59-73.

Naylor, T.H. (1971). Computer Simulation Experiments with Models of Economic Systems, Wiley, New York.

Naylor, T.H., J.L. Balintfy, D.S. Burdick, and K. Chu. (1967a). Computer Simulation Techniques, Wiley, New York.

Naylor, T.H., D.S. Burdick, and W.E. Sasser (1967b). "Computer simulation experiments with economic systems: the problem of experimental design," J. Amer. Stat. Assoc., 62, 1315-1337.

Naylor, T.H. and J.M. Finger (1967). "Verification of computer simulation models," Management Sci., 14, 92-101.

Naylor, T.H., K. Wertz, and T. Wonnacott (1968). "Some methods for evaluating the effects of economic policies using simulation experiments," Rev. Int. Stat. Inst., 36, 184-200.

Nelson, R.T. (1966). Systems, Models and Simulation, Mimeographed notes, Graduate School of Business Administration, University of California, Los Angeles, Calif.

Niedereichholz, J. (1971). "Business simulation and management decisions," Management Int. Rev., 11, 47-52.

Orcutt, G.H., M. Greenberger, J. Korbel, and A.M. Rivlin. (1961). Microanalysis of Socioeconomic Systems: A Simulation Study. Harper, New York.

Overholt, J.L. (1969). "The problem of factor selection," in The Design of Computer Simulation Experiments (T.H. Naylor, ed.), Duke University Press, Durham.

Parzen, E. (1962). Stochastic Processes. Holden-Day, San Francisco, Calif.

Pegels, C.C. (1969). "Simulation and the optimal design of a production process," Int. J. Prod. Res., 7, 219-231.

Saleeb, S. and M.G. Hartley (1968). "Simulation of traffic behaviour through a linked-pair of intersections," Transportation Res., 2, 51-61.

Schlaifer, R. (1959). Probability and Statistics for Business Decisions, McGraw-Hill, New York.

Schmidt, J.W. and R.E. Taylor (1970). Simulation and Analysis of Industrial Systems, Richard D. Irwin, Inc., Homewood Ill.

Schrank, W.E. and C.C. Holt (1967). "Critique of 'verification of computer simulation models'," Management Sci., 14, 104-106.

Smith, J.U.M. (1970). "Computer simulation of industrial operations," in Progress of Cybernetics, Volume 2 (J. Rose, ed.), Gordon and Breach, London.

Smith, W.P. and J.C. Shah (1964). "Design and development of a manufacturing systems simulator," J. Ind. Eng., 15, 214-220.

Starbuck, W.H. (1971). "Testing case-descriptive models," in Computer Simulation of Human Behavior (J.M. Dutton and W.H. Starbuck, eds.), Wiley, New York.

Stash, S.F. (1968). Some aspects of Multidimensional Verification of Simulations. Presented at the Symposium on the design of computer simulation experiments, Duke University, Durham, North Carolina, 14-16 Oct. 1968. (An abstract is published in The Design of Computer Simulation Experiments (T.H. Naylor, ed.), Duke University Press, Durham, N.C. 1969.)

Tocher, K.D. (1963). The Art of Simulation, English Universities Press, London.

Van Daal, J. and F. van Doeland (1972). Waiting for the Bridge, Report 7208, Econometric Institute, Netherlands School of Economics, Rotterdam.

Van Horn, R.L. (1971). "Validation of simulation results," Management Sci., 17, 247-258.

Wagner, H.M. (1971). "The ABC's of OR," Operations Res., 19, 1259-1281.

Wetherill, G.B. (1966). Sequential Methods in Statistics, Methuen, London, and Wiley, New York.

Chapter III

VARIANCE REDUCTION TECHNIQUES[1)]

III.1. INTRODUCTION AND SUMMARY

In this chapter we shall investigate some variance reduction techniques (VRT). In Section II.4 we mentioned that a VRT implies that "straight on" or "crude" sampling is replaced by more sophisticated sampling. Or more precisely, a VRT reduces the variance of the estimator by replacing the original sampling procedure by a new procedure that yields the same expected value but with a smaller variance. Some VRT's change the original sampling process completely. For instance, as we shall see in Section III.5 importance sampling means that the original distribution from which we sample is replaced by a completely new distribution. Other VRT's use the same sampling process as in crude sampling, but after sampling has ended, they do not use the simple average $\bar{x}$ but a more sophisticated estimator. Examples are stratification after sampling and control variates to be discussed in the Sections III.2.4 and III.4, respectively. Some VRT's modify the sampling process in a very subtle way as we shall see in Sections III.6 and III.7 where we shall discuss antithetic variates and common random numbers.

In the literature on VRT's attention is usually restricted to the estimation of the mean or the expected value. Hence we suppose that the purpose of the simulation experiment is the estimation of the mean value of the response of the simulated system. This response can be the waiting time of a customer in the steady state, the total profit of a firm over the planning period, etc. We shall not investigate if and how a VRT can be applied to improve the estimation of the variance, a quantile or a serial correlation coefficient.[2)]

We observe that the aim of some simulation studies is not the estimation of a mean but the estimation of the probability that the response is greater than some fixed quantity, i.e., a percentile. An example is a queuing system where we want to estimate the probability that the waiting time is greater than some fixed number, or an economic system for which we want to know the probability that the balance of payments deficit is larger than some fixed amount. However, it is very simple to formulate the estimation of such a probability as the estimation of a mean. For, as we have already seen in Eq. (2.5) of Appendix I.2 we can introduce a new variable, say $\underline{y}$, with the expected value p where p is the probability that the variable of interest, say $\underline{x}$, exceeds a fixed quantity.

In the literature the efficiency of a VRT is usually measured by the decrease of the variance of the estimator of the mean.[3] At first sight one might wonder why the variance is used instead of the standard deviation. For the reliability of an estimate can be measured by confidence limits of a form as in (1).

$$\bar{\underline{x}} \pm a(\sigma/\sqrt{n}) \tag{1}$$

where $\bar{\underline{x}}$ is the average of n independent observations, i.e., n independent simulation runs, σ is the standard deviation of an individual observation and a is suitable constant. Nevertheless it also makes sense to measure the reduction of the variance. As we shall see in Chapter V the number of observations, say n_w, needed to realize a certain reliability can be expressed as a function of the variance as in (2):

$$n_w = b\sigma^2 \tag{2}$$

where b is a suitable constant. Summarizing, in our opinion it is justified to measure the decrease of the standard deviation due to applying a VRT, if the sample size is supposed to be fixed since the relative change of the standard deviation is equal to that of the length of the confidence interval. If we suppose that the desired reliability is predetermined, then we should measure the variance since its relative decrease is equal to that of the number of observations needed to realize the prefixed reliability. Of course a variance reduction implies a standard deviation reduction, and vice versa. Note that we define the variance reduction (as a percentage) as $(\sigma_0^2 - \sigma_1^2)/\sigma_0^2$ ($\times$ 100%) where σ_0^2 and σ_1^2 are the old and new variances, respectively. Radema (1969) and Andréasson (1971b, pp. 13-14) discuss the _standard error_ of the estimated variance reductions. Raj (1968, p. 190) gives a formula for the coefficient of variation of a variance estimator, i.e., $\{\mathrm{var}(\hat{\underline{\sigma}}^2)\}^{1/2}/E(\hat{\underline{\sigma}}^2)$. Robust confidence intervals for variance estimators can be based on the "jackknife" technique (Gray and Schucany, 1972, pp. 176-182); the jackknife is described in Appendix 1 and will be used later in this Chapter. Besides the reduction of the standard deviation or variance we have to take into account the _extra computer time_ that a VRT may require. Unfortunately, this computer time heavily depends on the efficiency of the individual programmer and the type of computer. Nevertheless, we shall give the results of some VRT's corrected for this extra computer time. We also refer to Handscomb (1968, p. 2), Lewis (1972, p. 11), and Moy (1965, pp. 8-11); Handscomb calls a technique variance-reducing "if it reduces the variance proportionately more than it increases the work involved." Further, besides the computer running time we have to take into account the time it takes to construct the computer program. We shall see later on in this chapter that unlike some sophisticated VRT's, antithetic variates and common random numbers hardly take any _extra programming time_.

VRT's were originally applied not in simulation but in _Monte Carlo_ studies.[4)] (We still define Monte Carlo in the narrow sense of Section I.5.) A sophisticated survey of the many VRT's available for Monte Carlo work is given by Hammersley and Handscomb (1964). Later these VRT's were also applied in _simulation_. Tocher (1963, p. 171) suggested the use of VRT's in the simulation of complex industrial systems. Nevertheless, with Maxwell (1965, p. 2) and Moy (1965,

p. 21) we conclude that VRT's have hardly been applied in simulation; Marshall (1958) and Meier et al. (1969, pp. 274-275) discuss the reasons that may explain this fact. The application of a VRT in simulation raises some new problems compared with Monte Carlo studies. For the application of VRT's we would mention as relevant differences between simulation and Monte Carlo the following two aspects (which were also mentioned in Section I.6.2; for a different characterization we refer to Garman (1971, pp. 3-6) and (1973, pp. 5-7); see also Eqs. (178) through (173) below.)

(i) The observations in Monte Carlo studies are independent. In simulation, however, we experiment with the model over time so as a rule the observations are serially correlated. This difference is demonstrated by the examples in the appendices of Chapter I: In Appendix I.1 we estimated the value of the integral from the n observations $g(\underline{x}_i)$ as shown by Eq. (1.6) in Appendix I.1. Each observation $g(\underline{x}_i)$ is independent. In Appendix I.2 we estimated p (the probability that $\underline{x}$ is smaller than the fixed quantity a) from the N observations $\underline{y}_j$ as (2.10) in Appendix I.2 showed. Again these observations are independent. In the simulation example of Appendix I.3 the N trips on the same day are serially correlated. Nevertheless the trips on different days are independent for the strategies α and β. However, for strategy γ even the trips on different days are dependent. Note that in a queuing system the individual waiting times are dependent. So in general, in simulation studies the observations are serially correlated. (Obviously we can replicate a simulation run using a new sequence of random numbers; each run is then independent of the other runs.)

(ii) In Monte Carlo problems it is possible to express the response as a rather simple function of the stochastic input variables. In simulation the response is usually a very complicated function of the input and can be expressed explicitly only by the computer program itself; see also Garman (1971, p. 5) and (1973).

Because of these differences we have to adjust the VRT's that are used in Monte Carlo studies, such that they can be applied in simulation. For these adaptations we especially mention the little-known publications by Moy (1965) and (1966). Moy shows how stratified, importance and regression sampling can be so adjusted that they are applicable in the simulation of all types of systems. We shall further discuss the application in simulation of selective sampling, antithetic variates and common random numbers. As we shall see these six techniques are so devised that they can be applied in the simulation of complex systems. Numerous VRT's with limited applicability in simulation can be found in the literature.[5)]

So in the following sections we shall concentrate on the above six VRT's that can be used in the simulation of a wide range of systems. These VRT's can also be applied in Monte Carlo studies. Some VRT's, especially stratified and importance sampling, have a simpler form when applied to Monte Carlo problems. These simple forms will be presented before we discuss the VRT adjusted for application in simulation studies.

A description and a critical appraisal of the six techniques will be given. Extensions, limitations and corrections for the existing procedures will result. For example, we shall find that it is not correct to apply Moy's stratification procedure in simulations with a stochastic runlength; on the other hand, we shall extend stratified sampling in simulation to stratification after sampling. Brenner's selective sampling technique will be shown to yield a biased estimator and not to be suited for simulations with a stochastic runlength. The application of the jackknife statistic with control variates will be emphasized. We shall give a Monte Carlo example where a variance reduction of more than 99% is realized when applying importance sampling (a new density function being selected through minimization of the range of the estimator). Multiple responses will be found to lead to serious problems in importance sampling. It will be explained why we disagree with Tocher's "2^K factorial experiment" approach for antithetic variates, and several procedures will be presented for generating antithetic random numbers.

Further, it is customary in the literature to neglect the analysis of experiments in which variance reduction techniques are applied. Since this analysis is complicated by the use of these techniques we have also studied this aspect. The above VRT's, except for antithetic variates, will be found to give complications when deriving confidence limits.

Finally, we shall investigate the joint application of the two simplest techniques, antithetic variates and common random numbers. We shall see that due to undesirable correlation effects such a joint application may not be optimal. Therefore a procedure will be developed that yields the largest variance reduction; at the same time it allocates the available computer time in an efficient way. The practitioner who is interested only in straightforward VRT's, is advised to proceed directly to the Sections III.6 and III.7 dealing with antithetic variates and common random numbers.

III.2. STRATIFIED SAMPLING

III.2.1. Fundamentals of Stratified Sampling

Stratified sampling is an old procedure that is quite common in sample surveys. Therefore, before discussing stratification in Monte Carlo and simulation studies, we shall present the general principles of stratification. Suppose we want to estimate μ, the average consumption of some population. To estimate this population mean we take a sample of n individuals from that population. The well-known simple estimator of the mean μ is the average $\bar{\underline{x}}$ defined in (3).

$$\bar{\underline{x}} = \sum_{i=1}^{n} \underline{x}_i/n \tag{3}$$

where $\underline{x}_i$ is the consumption of the i'th (i = 1, ..., n) randomly drawn individual. As we know, $\bar{\underline{x}}$ is an unbiased estimator with variance σ^2/n where σ^2 is the variance of an individual observation $\underline{x}_i$.

In stratified sampling we measure besides the variable of interest $\underline{x}_i$ $(i = 1, \ldots, n)$, an extra variable, say $\underline{y}_i$, for each sampled individual i. This extra variable is called the stratification variable. It serves to classify each sampled individual into one of the, say, K classes or strata. These strata are formed by dividing the range of the stratification variable y into K nonoverlapping exhaustive classes. Denote the k'th stratum by S_k $(k = 1, \ldots, K)$. Suppose that in our example the stratification variable is income. Then an individual with consumption x belongs to class k if his income y falls into stratum S_k. In order to use stratification we shall need to know the probability that an individual belongs to the k'th class. This probability we denote by p_k. So we assume that we know[6])

$$p_k = P(\underline{y} \in S_k) \tag{4}$$

We may decide to observe n_k individuals with an income that falls into stratum k. Then we can estimate the population mean μ by the stratified estimator, say $\bar{\underline{x}}_{ST}$, defined in (5).

$$\bar{\underline{x}}_{ST} = \sum_{k=1}^{K} p_k \bar{\underline{x}}_k \tag{5}$$

where $\bar{\underline{x}}_k$ is the estimated mean of the k'th stratum as given by (6).

$$\bar{\underline{x}}_k = \sum_{j=1}^{n_k} \underline{x}_{kj}/n_k \tag{6}$$

with $\underline{x}_{kj}$ being the j'th $(j = 1, \ldots, n_k)$ observation in stratum k. The stratified estimator is unbiased as is proved in (7), where we use the definition in (5) and denote the mean of stratum k by μ_k.

$$E(\bar{\underline{x}}_{ST}) = \sum_{k=1}^{K} p_k E(\bar{\underline{x}}_k) = \sum_{k=1}^{K} p_k \mu_k = \mu \tag{7}$$

The last equality in (7) may be obvious to the reader; otherwise we refer to e.g., Fisz (1967, p. 84). The variance of the stratified estimator is derived in (8) using (5) and (6).

$$\operatorname{var}(\bar{\underline{x}}_{ST}) = \sum_{k=1}^{K} p_k^2 \operatorname{var}(\bar{\underline{x}}_k) = \sum_{k=1}^{K} p_k^2 \frac{\sigma_k^2}{n_k} \tag{8}$$

σ_k^2 being the variance within stratum k, i.e., σ_k^2 is the variance of $\underline{x}_{kj}$, as shown by (9).[7]

$$\sigma_k^2 = E[(\underline{x} - \mu_k)^2 | \underline{x} \in S_k] \tag{9}$$

Before investigating if stratification can lead to variance reduction, we shall see how we can establish confidence intervals for the mean μ. Obviously σ_k^2 can be estimated by $\underline{s}_k^2$ defined in (10).

$$\underline{s}_k^2 = \sum_{j=1}^{n_k} (\underline{x}_{kj} - \bar{\underline{x}}_k)^2/(n_k - 1) \tag{10}$$

This estimator can be substituted into (8) to yield the unbiased estimator of $\operatorname{var}(\bar{\underline{x}}_{ST})$ given in (11).

$$\underline{s}^2(\bar{\underline{x}}_{ST}) = \sum_{k=1}^{K} p_k^2 \frac{\underline{s}_k^2}{n_k} \tag{11}$$

Now we can follow Cochran (1966, pp. 94-95) and determine the familiar $(1 - \alpha)$ confidence interval for the population mean μ given in (12).

$$\bar{\underline{x}}_{ST} \pm z^{\alpha/2} s(\bar{\underline{x}}_{ST}) \tag{12}$$

where $\underline{s}(\bar{\underline{x}}_{ST})$ is the square root of $\underline{s}^2(\bar{\underline{x}}_{ST})$ in (11). The value of $z^{\alpha/2}$ can be found in the table for the normal distribution if we assume that $\bar{\underline{x}}_{ST}$ is normally distributed and $\underline{s}(\bar{\underline{x}}_{ST})$ is well determined. Concerning the normality assumption we note that we may assume that $\bar{\underline{x}}_k$, the stratum average, is asymptotically normal because of the central limit theorem. The weighted sum of independent normally distributed variables is also normal; compare Fisz (1967, p. 147, 150). Hence we may assume that $\bar{x}_{ST}$ is normally distributed, at least approximately. However, if we do not have many observations in each stratum, then the asymptotic normality may not hold and $s(\bar{x}_{ST})$ is not well determined. Then, following Cochran, we may determine $z^{\alpha/2}$ in (12) from the table for Student's t-statistic. We point out that Scheffé (1964, pp. 335, 346) found that the t-statistic is not very sensitive to violations of the normality assumption. A more serious problem is that $s^2(\bar{x}_{ST})$ has not the usual form $\Sigma(x_i - \bar{x})^2/(n - 1)$ but is a more complicated function as (11) showed. Cochran (1966, p. 94) suggests applying an approximation evolved by Satterthwaite (1946), see also Raj (1968, pp. 193-194). This approximation still determines $z^{\alpha/2}$ in (12) from Student's table, but uses a corrected number of degrees of freedom, say n_c, where n_c is given in (13).[8]

$$n_c = \frac{\{s^2(\bar{x}_{ST})\}^2}{\sum_{k=1}^{K} (p_k s_k)^4/(n_k^3 - n_k^2)} \tag{13}$$

As Cochran (1966, p. 95) further observes, Satterthwaite's approximation assumes that the individual observations x_{kj} are normal. Now it may be that we are interested in the average steady-state response of a system. From the stationary r-dependent central limit theorem given in Fraser (1957, p. 219) and in Chapter V it follows that we may assume the average response of such a stationary system to be asymptotically normal. However, if we are studying a nonstationary system, then this theorem does not apply, and the confidence interval of (12) becomes even cruder. Finally, the total sample size may not be fixed, but may depend on the variability of the estimator of the population mean, this variability being estimated during the sampling process. Then the number of observations per stratum becomes stochastic. This stochastic character complicates the derivation of confidence limits since limits as in (12) are based on a fixed sample size. (We shall return to this problem in Section III.8 and in Chapter V.) Summarizing, the confidence limits for the stratified estimator hold only approximately. (It might be possible to apply the jackknife technique to the stratified estimator but this requires additional research.)

Let us return to the variance of the stratified estimator $\bar{x}_{ST}$ as compared with that of the simple estimator $\bar{x}$. From (8) it follows that the variance of $\bar{x}_{ST}$ depends on the choice of n_k, the number of observations in stratum k. An obvious choice is to take 10% of the sample from stratum k if 10% of the population falls into stratum k. More generally, n_k is chosen in accordance with (14).

$$n_k = p_k n \tag{14}$$

If we choose n_k as in (14) we have so-called proportionally stratified sampling. Substituting (14) into (8) yields the variance of the proportionally stratified estimator $\bar{x}_{PS}$, given in (15).

$$\operatorname{var}(\bar{x}_{PS}) = \frac{1}{n} \sum_{k=1}^{K} p_k \sigma_k^2 \tag{15}$$

The relationship between the variance of the simple estimator $\bar{x}$ given in (3) and the variance of the proportionally stratified estimator $\bar{x}_{PS}$ is presented in (16); this relationship is proved for discrete variables by Cochran (1966, pp. 98-99) and for continuous variables by Tocher (1963, pp. 106-107).

$$\operatorname{var}(\bar{x}) = \operatorname{var}(\bar{x}_{PS}) + \sum_{k=1}^{K} p_k (\mu_k - \mu)^2 / n \tag{16}$$

Hence (16) shows that stratification is useful if the μ_k's, the means of the various strata, are not all equal (i.e., equal to μ). The means μ_k do differ if the variable of interest x depends on the stratification variable y. In our example we may indeed assume that x, consumption, is correlated with y, income. (In this example we have positive correlation since increasing income tends to increase consumption.) Consequently, stratification is efficient if there is a stratification variable that is strongly correlated with the variable of interest. Obviously, in order to be able to apply stratification we must know the stratum probabilities p_k, i.e., we must know the distribution of the stratification variable as (4) shows.

We can try to determine the optimum number of observations per stratum instead of using the proportional allocation given by (14) above. The optimum allocation can be found in Cochran (1966, pp. 95-97), either taking into account different sampling cost per stratum or not. However, this optimum allocation can be realized only if we know the variances within each stratum, i.e., σ_k^2 defined in (9) above. Since usually the σ_k^2's are unknown we will not further discuss optimum allocation.

The above discussion of stratification can also be found in the handbook on stratification written by Cochran (1966) and in the studies on VRT's by Moy (1965) and (1966). We refer to Cochran (1966, pp. 87-153) for a discussion of the selection of the optimum number of observations per stratum in case there is more than one response variable of interest; the use of two stratification variables instead of a single variable; the optimum choice of the stratum boundaries and the number of strata; the estimation of the efficiency of stratification from the sample; the estimation of the variance if some stratum contains only one observation so the denominator of (10) would become zero, etc.; see also Henzler (1970) and Raj (1968, pp. 61-84, 151-152, 206-209). We further mention the exposé by Moy (1965, pp. 83-84) showing that the number of strata should be small, say between 3 and 10. Next we shall present a stratification variant that is only briefly discussed by Cochran (1966, pp. 135-136) and is not mentioned at all by Moy. Nevertheless, we shall see in Section III.2.4 that this variant, called stratification after sampling, may very well be important for simulation.

III.2.2. Stratification after Sampling

Suppose that the sampling process consists of taking a simple, i.e., a nonstratified, sample of n observations. However, instead of the simple sample average $\underline{\bar{x}}$ given in (3) above, we stratify this sample afterward. Unlike proportionally stratified sampling, the number of observations in each stratum is not prefixed, but depends on the sample outcome. This new estimator, say $\underline{\bar{x}}_{SA}$, is defined in (17).

$$\underline{\bar{x}}_{SA} = \sum_{k=1}^{K} p_k \underline{\bar{z}}_k \tag{17}$$

where

$$\underline{\bar{z}}_k = \sum_{j=1}^{\underline{n}_k} \underline{x}_{kj}/\underline{n}_k \tag{18}$$

and

$$\sum_{k=1}^{K} \underline{n}_k = n \tag{19}$$

We first compare the estimator $\underline{\bar{x}}_{SA}$ with the simple sample average $\underline{\bar{x}}$. We rewrite $\underline{\bar{x}}$, defined in (3), as in (20).

$$\begin{aligned} \underline{\bar{x}} &= n^{-1} \sum_{i=1}^{n} \underline{x}_i = n^{-1} \sum_{k=1}^{K} \sum_{j=1}^{\underline{n}_k} \underline{x}_{kj} \\ &= \sum_{k=1}^{K} \frac{\underline{n}_k}{n} \sum_{j=1}^{\underline{n}_k} \frac{\underline{x}_{kj}}{\underline{n}_k} \\ &= \sum_{k=1}^{K} \frac{\underline{n}_k}{n} \underline{\bar{z}}_k \end{aligned} \tag{20}$$

where $\underline{\bar{z}}_k$ is defined in (18) above. Comparison of (20) with (17) shows that in stratification after sampling the empirical weights $\underline{n}_k/n$ are replaced by the "true" or theoretical weights p_k.

Next we shall consider the usual stratified estimator with a prefixed number of observations per stratum. This estimator was defined in (5) and (6) above. In the case of proportional stratification the prefixed number of observations per stratum has the particular value $(p_k n)$ as (14) specified. Comparison of (18) with (6) shows that in stratification after sampling the number of observations per stratum is no longer fixed but depends on the sample results and hence it has become a stochastic variable.[9)]

Cochran (1966, pp. 135-136) showed that stratification after sampling is nearly as precise as proportionally stratified sampling, provided that the samples per stratum are reasonably large, say > 20. We observe that intuitively it is reasonable that the efficiency of stratification after sampling approaches that of proportional stratification, since the expected value of the stochastic subsample size $\underline{n}_k$ is equal to the prefixed subsample size in proportional sampling. For in stratification after sampling we have

$$E(\underline{n}_k) = p_k n \tag{21}$$

and this can be compared with the subsample size for proportional stratification given in (14). Stratification after sampling is attractive if it is impossible or difficult to fix the number of observations per stratum; therefore the stratum to which an observation belongs, is known only after the actual sampling.

As we have just seen Cochran's statement about the efficiency of stratification after sampling depends on the condition that the subsamples are "reasonably large, say > 20." This condition will not be very restrictive for broad strata (so p_k is not very small) and a large total sample size n. Otherwise we have to reckon with the possibility that the stochastic variable $\underline{n}_k$ is small and may even be zero. If $\underline{n}_k = 0$, i.e., the k'th stratum remains empty, then we cannot apply the formula for $\bar{\underline{x}}_{SA}$, since (18) shows that we would divide by zero. Therefore, if we take into account that $\underline{n}_k$ may be zero, then we have to correct the definition of stratification after sampling given in (17) through (19). One possible correction is suggested by Cochran (1966, p. 136), but is not worked out by him. He proposes combining two or more strata if one or more $\underline{n}_k$ is zero. We formalize Cochran's suggestion in (22) which gives the definition of $\bar{\underline{x}}_{SAC}$, the estimator based on stratification after sampling with <u>combination</u> of strata if necessary.

$$\bar{\underline{x}}_{SAC} = \sum_{k=1}^{K} p_k \bar{\underline{w}}_k \tag{22}$$

with

$$\bar{\underline{w}}_k = \sum_{j=1}^{\underline{n}_k} \underline{x}_{kj}/\underline{n}_k \quad \text{if} \quad \underline{n}_k > 0 \tag{23a}$$

$$\bar{\underline{w}}_k = \bar{\underline{w}}_g = \sum_{j=1}^{\underline{n}_g} \underline{x}_{gj}/\underline{n}_g \quad \text{if} \quad \underline{n}_k = 0, \quad \underline{n}_g > 0 \tag{23b}$$

and

$$\sum_{k=1}^{K} \underline{n}_k = n \tag{24}$$

In Appendix III.2 we derive that in general Cochran's suggestion leads to a <u>biased</u> estimator. If this bias and the variance of $\bar{\underline{x}}_{SAC}$ is small, then $\bar{\underline{x}}_{SAC}$ is still an attractive estimator. Bias and variance together are measured by the <u>mean square error</u>, or briefly the MSE. Let us briefly review the MSE before proceeding.

The MSE as defined by Goldberger (1964, p. 126) is the expected value of the squared deviation between the estimator and the true value of the population parameter that the estimator is supposed to estimate.[10)] In our case the estimator is $\bar{\underline{x}}_{SAC}$ which should estimate μ, and hence the MSE is given by (25).

$$\text{MSE}(\bar{\underline{x}}_{SAC}) = E[(\bar{\underline{x}}_{SAC} - \mu)^2] \tag{25}$$

In Appendix III.3 we prove a well-known relation, viz., the MSE is equal to the <u>variance</u> of the estimator plus the square of the <u>bias</u> of the estimator. So in our case (26) holds.

$$\mathrm{MSE}(\underline{\bar{x}}_{SAC}) = \mathrm{var}(\underline{\bar{x}}_{SAC}) + [E(\underline{\bar{x}}_{SAC}) - \mu]^2 \tag{26}$$

where

$$\mathrm{var}(\underline{\bar{x}}_{SAC}) = E[\{\underline{\bar{x}}_{SAC} - E(\underline{\bar{x}}_{SAC})\}^2] \tag{27}$$

Further we derive in Appendix III.3 that an unbiased estimator of the MSE of $\underline{\bar{x}}_{SAC}$ can be calculated if we have N replicated samples so we have N values of the estimator $\underline{\bar{x}}_{SAC}$, say $\underline{\bar{x}}_{SACi}$ $(i = 1, \ldots, N)$, and if we know the value of the population parameter μ. (In Monte Carlo studies of the efficiency of various estimators we do know μ.) The unbiased estimator of the MSE in (25) is given by (28).

$$\widehat{\underline{\mathrm{MSE}}}(\underline{\bar{x}}_{SAC}) = \sum_{i=1}^{N} (\underline{\bar{x}}_{SACi} - \mu)^2/N \tag{28}$$

It may be possible to estimate--at least approximately--the MSE even if the population mean is unknown; see Raj (1968, pp. 88-93, 105, 149). After this brief exposé on the MSE we return to the biased estimator $\underline{\bar{x}}_{SAC}$.

Since the MSE measures the variance and the bias jointly, it is a useful criterion for choosing among estimators if one or more estimators may be biased. (If some estimator is unbiased, then its MSE reduces to its variance.) Unfortunately, it does not seem easy to evaluate the MSE of $\underline{\bar{x}}_{SAC}$ (see also footnote 11). Notice that the bias of $\underline{\bar{x}}_{SAC}$ depends on the particular population being sampled. For this population determines the bias that is created by replacing $\underline{\bar{w}}_k$ by $\underline{\bar{w}}_g$ in (22).

Next we shall consider a second way to correct the estimator based on stratification after sampling if we reckon with the possibility that a stratum remains empty. Tocher (1963, p. 105) suggests that sampling be continued until each stratum contains at least one observation. In this case the total sample size cannot be fixed equal to n. For as long as a stratum is empty, we have to continue (simple) sampling. We have formalized Tocher's suggestion in (29) through (32). These equations define $\underline{\bar{x}}_{SAP}$, the estimator based on stratification after sampling with prolonged sampling until each stratum is nonempty.

$$\underline{\bar{x}}_{SAP} = \sum_{k=1}^{K} p_k \underline{\bar{v}}_k \tag{29}$$

where

$$\underline{\bar{v}}_k = \sum_{j=1}^{\underline{n}_k} \underline{x}_{kj} / \underline{n}_k \tag{30}$$

$$\underline{n}_k > 0 \quad \text{for all} \quad k \tag{31}$$

$$\sum_{k=1}^{K} \underline{n}_k = \underline{n} \tag{32}$$

It is simple to prove that this estimator is <u>unbiased</u>. It seems difficult to compare the variance of $\underline{\bar{x}}_{SAP}$ with the MSE of $\underline{\bar{x}}_{SAC}$ or with the variance of $\underline{\bar{x}}_{PS}$ and $\underline{\bar{x}}$. More research on this topic is necessary; see also Williams (1964, pp. 1060-1061).

We saw above how we can determine the <u>approximate confidence limits</u> (12) for the population mean when applying stratified sampling with a <u>fixed</u> number of observations per stratum. When we stratify <u>after</u> sampling, these limits may become even cruder. For in the definition of the stratum averages $\underline{\bar{z}}_k$, $\underline{\bar{w}}_k$ and $\underline{\bar{v}}_k$ of the estimators $\underline{\bar{x}}_{SA}$, $\underline{\bar{x}}_{SAC}$, and $\underline{\bar{x}}_{SAP}$, respectively the number of observations per stratum, $\underline{n}_k$, is <u>stochastic</u>. The estimated standard deviation to be used in the confidence interval (12) can be derived as follows. In Appendix III.4 we prove that an unbiased estimator of the variance of $\underline{\bar{v}}_k$ is given by (33).

$$\widehat{\underline{\text{var}}}(\underline{\bar{v}}_k) = \frac{\sum_{j=1}^{\underline{n}_k} (\underline{x}_{kj} - \underline{\bar{v}}_k)^2}{(\underline{n}_k - 1)} \frac{1}{\underline{n}_k} \tag{33}$$

where we replace the original condition (31) by (34) in order to make the denominator of (33) nonzero:

$$\underline{n}_k > 1 \quad \text{for all} \quad k \tag{34}$$

Hence an unbiased estimator of the variance of the estimator $\bar{\underline{x}}_{SAP}$ is given by (35).

$$\underline{\hat{var}}(\bar{\underline{x}}_{SAP}) = \sum_{k=1}^{K} p_k^2 \; \underline{\hat{var}}(\bar{\underline{v}}_k) \tag{35}$$

There is a completely analogous estimator for the variance of $\bar{\underline{x}}_{SA}$ assuming that $\underline{n}_k > 1$ in (18). The estimator $\bar{\underline{x}}_{SAC}$ raises more problems since it is a biased estimator and it seems hard to derive an unbiased estimator for its variance.[11] So the determination of confidence limits for the population mean μ is even more difficult than with the other stratified estimators.

We finally observe that a simple expedient for realizing a non-stochastic number of observations can be found in Handscomb (1968, p. 6) and is as follows. Determine the desired number of observations per stratum, take one observation, and determine to which stratum it belongs. If that stratum has been filled already, then reject this observation; otherwise accept it, etc. This rejection procedure will imply much wasted time if it takes long to sample an observation and to determine to which stratum it belongs. The difference with stratification as discussed in Section III.2.1 is that we assumed there that we know beforehand to which stratum an observation will belong. We shall see in Section III.2.4 how such knowledge can be obtained in simulation.

III.2.3. Stratified Sampling in Monte Carlo Studies

In his well-known book on simulation Tocher (1963) gives some examples of stratified sampling. Actually his examples do not concern simulation but Monte Carlo studies, more precisely distribution sampling which we discussed in Section I.5. Tocher (1963, p. 100) considers the estimation of the expected range. As we know the range $\underline{R}$ of a sample of m observations $\underline{x}_i$ $(i = 1, \ldots, m)$ is defined by (36).

$$\underline{R} = \max_i \underline{x}_i - \min_i \underline{x}_i \tag{36}$$

As a possible stratification variable Tocher (1963, p. 103) mentions the sample mean $\bar{\underline{x}}$ defined in (37).

$$\bar{\underline{x}} = \sum_{i=1}^{m} \underline{x}_i / m \tag{37}$$

In the Monte Carlo experiment we sample m values of $\underline{x}$ from some given distribution. These m observations give an estimate of the range and also an estimate of the stratification variable, the sample mean. Repeating such sampling n times gives the estimators $\underline{R}_1, \ldots, \underline{R}_n$ and $\bar{\underline{x}}_1, \ldots, \bar{\underline{x}}_n$. Next we form two strata, S_1 and S_2, such that $\bar{\underline{x}}_j$ (and therefore $\underline{R}_j$) $(j = 1, \ldots, n)$ falls into S_1 if $\bar{\underline{x}}_j \leq \mu$ and $\bar{\underline{x}}_j$ falls into S_2 if $\bar{\underline{x}}_j > \mu$, where μ is the known (population) mean of the distribution from which we sample. If we know that the density function from which we sample is symmetric, then we know the stratum weights p_k $(k = 1, 2)$. For Tocher (1963, p. 103) states that if there is no lump probability at $\underline{x} = \mu$ then the symmetry of the density function of $\underline{x}$ implies that of $\bar{\underline{x}}$ and consequently

$$p_1 = P(\bar{\underline{x}}_j \in S_1) = P(\bar{\underline{x}}_j \leq \mu) = \frac{1}{2} \tag{38}$$

and

$$p_2 = P(\bar{\underline{x}}_j \in S_2) = P(\bar{\underline{x}}_j > \mu) = \frac{1}{2} \tag{39}$$

So a stratified estimator of the expected range of m observations sampled from some given symmetric distribution, is shown in (40).

$$\bar{\underline{R}} = \sum_{k=1}^{2} p_k \bar{\underline{R}}_k$$

$$= \frac{1}{2} \sum_{g=1}^{\underline{n}_1} \underline{R}_{1g} / \underline{n}_1 + \frac{1}{2} \sum_{h=1}^{\underline{n}_2} \underline{R}_{2h} / \underline{n}_2 \tag{40}$$

where $\underline{R}_{1g}$ and $\underline{R}_{2h}$ denote a range with a corresponding sample mean that falls into strata 1 and 2, respectively. That the sample mean $\bar{\underline{x}}$ can serve as a stratification variable since it is correlated with the range $\underline{R}$ can be seen as follows. If some observation x_i among the m observations happens to be extremely high, then both $\bar{x}$ and R are higher than their respective true means, i.e., $\bar{\underline{x}}$ and $\underline{R}$ are positively correlated. Tocher (1963, pp. 104-105) further discusses other stratification variables that can be used when estimating the expected range.

Hillier and Lieberman (1968, pp. 455-458) also mention the use of stratification in simulation. Nevertheless their example actually concerns the Monte Carlo estimation of the mean of an exponential distribution. The variant of stratification they used for the evaluation of the resulting integral can be found in Hammersley and Handscomb (1964, pp. 55-57); see also Gaver (1969), Molenaar (1968, p. 120), Newman and Odell (1971, pp. 57-62), and our footnote 13 later on.

III.2.4. Stratified Sampling in Simulation

The stratification variable in the Monte Carlo study of the sample range discussed in the preceding section, was the sample mean. Both statistics are correlated when generated from the same random numbers. In the same way we could generate two related responses in simulation. As an example we consider the maintenance problem of Appendix I.3. Suppose that the probability function of the response for strategy β can be determined analytically, and that the mean response for strategy γ is to be estimated by simulation. Then we can simulate both strategy β and γ, even though β is supposed to be known. If both strategies are simulated with the same random numbers, then both responses are correlated. (We shall return to this correlation aspect in Section III.7 where we shall discuss the use of the same random numbers when simulating several systems.) Since strategy β also has an analytical solution we can calculate the stratum weights p_k, and use the responses of strategy β as a stratification variable. This procedure has two obvious drawbacks:

(i) There must be a related system with a known analytical solution.

(ii) It may take much extra time to program and run the computer program for the related system that yields the stratification variable. (In the Monte Carlo example of the previous section the extra time needed for the calculation of the stratification variable $\bar{x}$ is small compared with the time needed for generating x_i and calculating R.)[12]

Ehrenfeld and Ben-Tuvia (1962, pp. 261-263, 264-265) devised another stratification procedure for simple queuing systems. Unfortunately their procedure does not seem suited for general use in simulation; also compare Clark (1959), Price (1972) and Ringer (1965). So stratification can be rather simply employed in Monte Carlo studies, but we agree with Moy (1965, p. 85) that the procedures given by several authors for stratification in simulation all miss general applicability. Hence it is not surprising that Tocher (1963, p. 171) doubts whether stratification can be used at all in the simulation of complex systems. Nevertheless Moy (1965) succeeded in deriving a procedure that may have wide applicability in simulation. His stratification variable is not the response of a related system but is defined directly on the random numbers used to simulate the system of interest.[13] We shall next investigate his procedure.

Moy (1965, p. 90) and (1966, p. 17) proposes using as a stratification variable $\underline{y}$ defined in (41) and (42):

$$\underline{y} = \sum_{i=1}^{m} \underline{t}_i \tag{41}$$

with

$$\begin{aligned} \underline{t}_i &= 1 \quad \text{if } \underline{r}_i \geq c \\ &= 0 \quad \text{if } \underline{r}_i < c \end{aligned} \tag{42}$$

where c is a constant satisfying $0 < c < 1$ and $\underline{r}_1$ is the i'th $(i = 1, \ldots, m)$ random number in the sequence of random numbers used to generate one value of the response variable $\underline{x}$. Notice that Moy assumes that m, the total number of random numbers in one simulation run, is a known constant. The latter condition is met if e.g., we decide to simulate a single-channel queuing system for, say, m_1 customers; each customer needs two random numbers, one number for his arrival time and one number for his service time; hence $m = 2m_1$ in this example. We shall return to the condition of a known constant m later on in this section.

From (16) in Section III.2.1 it followed that a good stratification variable $\underline{y}$ should be correlated with the variable of interest $\underline{x}$. As Moy (1965, p. 90) and (1966, pp. 5, 17-18) remarks such a correlation can be created by designing the simulation model of a queuing system in such a way that congestion increases as the magnitude of the random number $\underline{r}_i$ increases. So in the above example of a simple queuing system we can generate exponential interarrival times, say $\underline{v}$, and exponential service times, say $\underline{w}$, using (43) and (44), respectively. (These equations are based on (18) and (20) in Section I.6.4.)

$$\underline{v}_i = -\lambda_v^{-1} \ln(\underline{r}_{2i-1}) = \lambda_v^{-1} |\ln(\underline{r}_{2i-1})| \tag{43}$$

and

$$\underline{w}_i = -\lambda_w^{-1} \ln(1-\underline{r}_{2i}) = \lambda_w^{-1} |\ln(1-\underline{r}_{2i})| \tag{44}$$

$$(i = 1, 2, \ldots, m_1)$$

where λ_v and λ_w are the parameters of the exponential distribution of $\underline{v}$ and $\underline{w}$, respectively, this distribution being specified in (15) of Section I.6.4. Note further that the random number $\underline{r}$ satisfies $0 \leq r \leq 1$ and therefore $\ln(r)$ and $\ln(1 - r)$ are negative. Hence a high value of a random number means that the interarrival time v is small or the service time w is high. This in turn leads to increased congestion, i.e., to a high value of the waiting time, which is supposed to be the variable of interest $\underline{x}$. Finally, from the definition of the stratification variable in (41) and (42) it follows that high values of $\underline{r}$ lead to a high value of the stratification variable $\underline{y}$. So $\underline{y}$ and $\underline{x}$ are indeed (positively) correlated. It is easy to see that (positive) correlation is also created between the stratification variable $\underline{y}$ of (41) and the response of the maintenance system described in Appendix I.3. We think that in models of other types of systems such correlation can be created too, so $\underline{y}$ in (41) can indeed be used as a stratification variable in the simulation of many systems.

A second condition for a stratification variable is that we know the stratum weights p_k defined in (4) above. The weights p_k can be calculated for this particular stratification variable as follows. From the definition of $\underline{y}$ in (41) and (42) we see that $\underline{y}$ is a binomially distributed variable with "probability of success" equal to $(1 - c)$. For

$$P(\underline{t}_i = 1) = P(\underline{r}_i \geq c) = 1 - c \tag{45}$$

and, using Fisz (1967, p. 136), it follows that the probability function of $\underline{y}$ is given by (46).

$$P(\underline{y} = j) = \binom{m}{j} (1 - c)^j c^{m-j} \quad (j = 0, 1, \ldots, m) \tag{46}$$

We can form K strata by dividing the range of $\underline{y}$ into K non-overlapping exhaustive intervals. For example, if $m = 500$ and $k = 5$, we may form the following strata S_k.[14)]

$$[0, 100), [100, 200), [200, 300), [300, 400), [400, 500] \tag{47}$$

Hence the stratum weight, say, p_2 is given by (48).

$$p_2 = P(\underline{y} \in S_2) = P(100 \leq \underline{y} < 200) = \sum_{k=100}^{199} P(\underline{y} = h) \tag{48}$$

where $P(\underline{y} = h)$ is given by (46).

Moy applies proportionally stratified sampling. (Later on we shall consider stratification after sampling using Moy's stratification variable.) From (14) it follows that in proportional stratification the number of observations in stratum k is given by (49).

$$n_k = p_k n \tag{49}$$

where n is the total number of observations, i.e., the total number of simulation runs (each run being of "length" m). Note that n is assumed to be a known constant. (In Section III.8.5 and Chapter V we shall return to this assumption since actually the number of runs may depend on the available computer time and the desired reliability.) From n, the known constant, and p_k, calculated from (48), we can determine n_k using (49).

Once we have calculated n_k, the number of observations in the k'th stratum, we have to sample n_k observations from that stratum in a random way. An observation belongs to stratum k if the simulation run yields a value of the stratification variable $\underline{y}$ that falls into stratum k. Moy devised the following procedure.

(i) Sample randomly a value of $\underline{y}$ from stratum k as follows. Use the definition of the conditional probability presented by Fisz (1967, p. 20) and repeated in (50).

$$P(A|B) = \frac{P(AB)}{P(B)} \quad \text{for} \quad P(B) > 0 \tag{50}$$

i.e., the probability of event A under the condition that B holds, is equal to the probability of the product AB divided by the probability of event B. So if we want to sample a value of $\underline{y}$ randomly from stratum 2 then we should sample $\underline{y}$ from the probability function given in (51).

$$P(\underline{y} = g | \underline{y} \in S_2) = \frac{p(\underline{y} = g \text{ and } \underline{y} \in S_2)}{P(\underline{y} \in S_2)}$$

$$= \frac{P(\underline{y} = g)}{P(\underline{y} \in S_2)} \qquad (g = 100, 101, \ldots, 199) \tag{51}$$

Substituting (46) and (48) into (51) yields (52).

$$P(\underline{y} = g | \underline{y} \in S_2) = \frac{\binom{m}{g} (1 - c)^g c^{m-g}}{\sum_{h=100}^{199} \binom{m}{h} (1 - c)^h c^{m-h}} \tag{52}$$

Generating $\underline{y}$ from (52) can be done using the inversion technique specified by (33) in Section I.6.4. A flowchart for the computer program is given in Fig. 1 which is based on Moy (1965, p. 91, Fig. 6-1). In Fig. 1, P(G) denotes $P(\underline{y} = g)$ as specified in (46) and W(2) is the stratum weight p_2 given in (48).

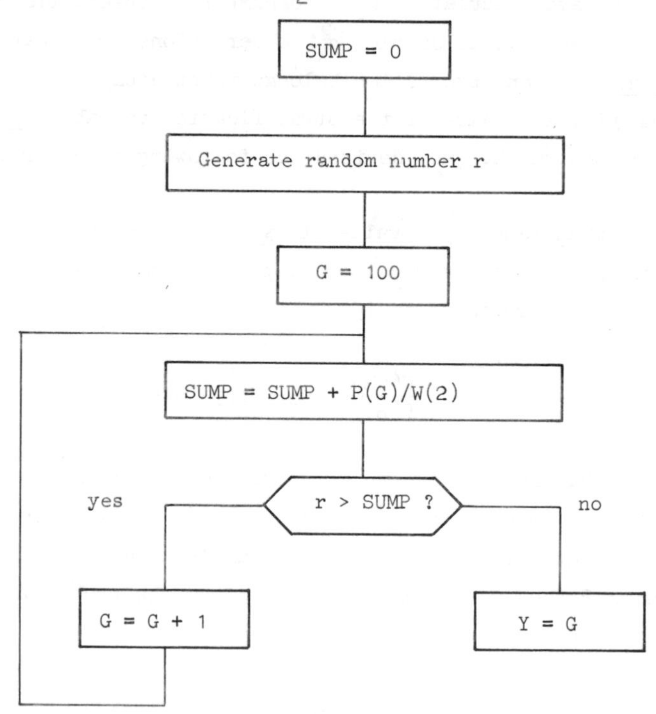

Fig. 1. Generating a value of the stratification variable $\underline{y}$ from stratum 2.

(ii) Once we have sampled a value, say y, of the stratification variable we have to generate an observation, i.e., a simulation run, which indeed yields that value y for the stratification variable. Suppose that we sampled the value $y = 175$ from stratum 2 using (52) or its flowchart equivalent in Fig. 1. From the definition of $\underline{y}$ in (41) it follows that 175 of the 500 ($= m$) random numbers should be not smaller than c. There are many, namely, $\binom{500}{175}$, possible vectors consisting of 500 random numbers, 175 numbers being not smaller than c. We should sample one of these vectors in a random way. In Fig. 2 the flowchart for such sampling is given. This flowchart is taken from Moy[15] (1965, p. 93, Fig. 6-2). Note that we use Fig. 2 every time we need a random number, i.e., 500 times in a simulation run; at the beginning of each run $M = 500$

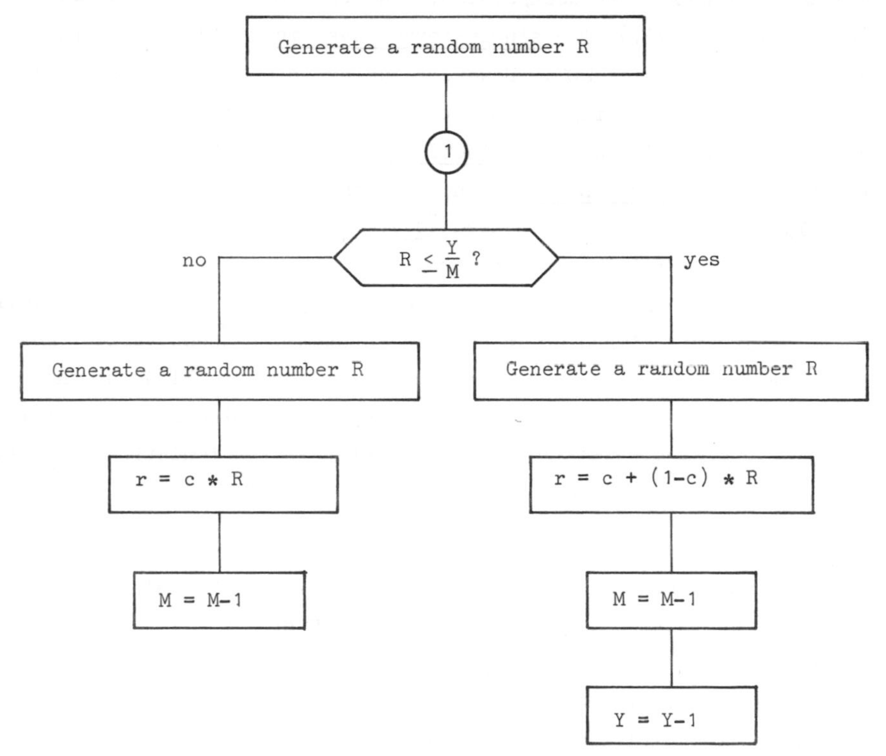

Fig. 2. Generating a random number $\underline{r}$ such that the stratification variable has the desired value.

and Y is equal to the value we sampled using Fig. 1. (We sample n values for the stratification variable using Fig. 1 since we replicate n times.) We observe that Fig. 2 amounts to sampling without replacement, as the probability of sampling a particular event changes during the sampling experiment. For in Fig. 2 the probability that a random number in block 1 is smaller than Y/M changes during the simulation run since Y and M change. The result is that the number of random numbers not smaller than c is exactly equal to the value we sampled for the stratification variable y. In Appendix III.6 we give a simple example to demonstrate the rationale of the procedure described by Fig. 2.

Moy (1965, pp. 97, 103, 104) and (1966, pp. 29, 32) applied proportionally stratified sampling to single-server queuing systems and to a complex truck dock system. (He put the constant c in (42) equal to 1/2.) The single-server systems yielded an average variance reduction of 20%. However, taking into account the extra computer time needed for the stratification, Moy (1965, p. 101) concludes that stratification is inefficient for this type of systems. For the complex truck system even a variance increases resulted! Moy (1965, pp. 103-104) also used "system-dependent" stratification, i.e., in the definition of the stratification variable y only the random numbers used to generate a special type of system event (e.g., the interarrival times) are used. The system-dependent variant gave better results for the truck dock system, but was still worse than the system-independent variant for the simple queuing system. Before deciding whether stratification is inefficient in simulation studies, we want to make two remarks about Moy's results.[16)]

(i) In the truck dock system m, the number of random numbers per simulation run, may vary from run to run. So we would replace m in the definition of the stratification variable given in (41) by m, where m denotes the stochastic number of random numbers needed in a run. Moy (1965, pp. 101-102) and (1966, p. 18) tried to solve this complication by assuming "that the length of the random number vector is equal to the length of the largest of the variable length vectors needed in the simulation. The generation of a system observation then uses only as many of these random numbers as are needed, and neglects the rest." In our opinion, however, this procedure is incorrect and we even think that stratification is <u>impossible</u> if there is not a constant number of random numbers per run. For the derivation of the stratum weights in (48) is based on the binomial distribution where m is a constant as (46) showed; sampling a value of the stratification variable is again based on a constant m as (52) showed. We shall further demonstrate our objections against Moy's procedure by an example. Suppose that the highest value of $\underline{m}$ likely to be required is 750. So we divide the range of the stratification variable into the following five strata S_k.

$$[0, 150),\ [150, 300),\ [300, 450),\ [450, 600),\ [600\text{-}750] \tag{53}$$

Suppose that using Fig. 1 we sample a value for the stratification variable from the last stratum equal to, say, 723. Next using Fig. 2, we start generating a vector of random numbers with the required composition, i.e., a vector of 750 random numbers, 723 numbers being not smaller than c. However, after, say, 500 random numbers we find that the simulation run stops. The value of $\underline{y}$ will then be approximately

$$723 \times (500/750) = 482 \tag{54}$$

Then this simulation run belongs to stratum 4 instead of 5! It might look reasonable to place the response $\underline{x}$ in stratum 4 and to use the stratum weight

$$p_4 = P(\underline{y} \in S_4 | m = 500) \tag{55}$$

But actually $\underline{x}$ was generated using the value 723 of $\underline{y}$, this value being sampled from (56), the analog of (52).

$$P(\underline{y} = g | \underline{y} \in S_5 \text{ and } m = 750) = \frac{P(\underline{y} = g)}{P(\underline{y} \in S_5)} \tag{56}$$

$$(g = 600, 601, \ldots, 750)$$

Moy (1965) and (1966) does not show how he solved this problem, i.e., how he determined in which stratum a response should be placed and which stratum weights should be used. Additional research is needed to check whether the bad results Moy found for the truck dock system are caused by an incorrect stratification procedure, or by the complexity of the system (which may disturb the desirable high correlation between the stratification variable and the response variable). For it is conceivable that in a complex system with a constant m, we find a variance reduction as in the simple queuing systems with a constant m. (We remark that a constant m in a complex system can be easily realized in steady-state simulations. For, as we saw in Section II.8, the length of the run can be fixed by the experimenter and therefore we can decide to stop the run when the number of random numbers used in the run is equal to the constant m. On the other hand, steady-state systems may be less suited for stratification since we may tend to increase the sample size by continuing the simulation run instead of replicating the run, and this may conflict with the requirement that no stratum be empty.)

(ii) We saw above that because of the extra computer time Moy rejected proportionally stratified sampling even for the simple queuing systems with a constant m. Therefore we suggest that an alternative stratification procedure be considered, namely, stratification after sampling. As we explained in Section III.2.2 in the latter procedure we do not fix the number of observations per stratum but instead we generate simulation runs, in the traditional way, each run using m random numbers. Afterward, applying the definition of the stratification variable $\underline{y}$ in (41), we determine to which stratum each run belongs and calculate the stratified estimator given in (17), or more precisely either the estimator given in (22) or that in (29). We saw in Section III.2.2 that the variance, or more generaly the mean square error, of this estimator is larger than that of the proportionally stratified estimator. However, no extra computer time is needed for generating a value of the stratification variable from Fig. 1 and for generating a vector of random numbers with the required composition from Fig. 2. In Section III. 2.2 we remarked that with stratification after sampling it seems more difficult to determine confidence limits for the population mean than with a fixed number of observations per stratum. Even in the latter case, however, the discussion of Eq. (12) showed that exact confidence limits cannot be established. So the determination of appropriate confidence limits in stratification needs more research.

III.3. SELECTIVE SAMPLING OR THE FIXED SEQUENCE METHOD

III.3.1. Description of the Method

We shall now consider a VRT devised by Brenner (1963). In the preceding sections we saw that stratification (except for stratification after sampling) changes the sampling process in such a way that the number of simulation runs with a value of the stratification variable $\underline{y}$ in a particular stratum k is fixed, namely, fixed to be n_k. This means that the number of simulation runs with a particular pattern of events is fixed. This is realized by sampling without replacement as Fig. 2 showed. Now Brenner fixes the pattern of events within a simulation run, since he realizes that all n runs have the same pattern. The pattern that is realized is such that the number of events of a particular type is equal to the theoretically expected number. This is further explained in the following description of his procedure.

Brenner implicitly assumes that there is only one kind of stochastic input variable. In the systems which Brenner (1963, pp. 293-294) considered this input is the demand for the article that is stocked in his inventory systems. Further this input variable, say $\underline{y}$, is a discrete variable and therefore we can distinguish, say, M possible values of that variable, as (57) shows.[17]

$$P(\underline{y} = y_i) = q_i \qquad (i = 1, 2, \ldots, M) \tag{57}$$

If we need m values of $\underline{y}$ in a simulation run, then we denote the number of times $\underline{y}$ has the value y_i by $\underline{v}_i$. Obviously the expected value of $\underline{v}_i$ is given by (58).

$$E(\underline{v}_i) = q_i m \tag{58}$$

Brenner <u>fixes</u> the number of times that a particular value of $\underline{y}_i$ is sampled in a run, i.e., he makes (59) hold.

$$\underline{v}_i = v_i \tag{59}$$

He wants to make v_i equal to the <u>expected</u> value of $\underline{v}_i$, i.e.,

$$v_i = q_i m \tag{60}$$

where

$$\sum_1^M v_i = m \sum_1^M q_i = m \tag{61}$$

In general, however, $q_i m$ is not an <u>integer</u>, whereas v_i has to be integer. Therefore Brenner (1963, p. 292, 925) makes $q_i m$ integer in such a way that the sum of the squared deviations between the "theoretical" values $q_i m$ and the integers v_i is minimized, i.e., he minimizes z defined in (62).

$$z = \sum_{i=1}^{M} (v_i - q_i m)^2 \tag{62}$$

subject to the condition given in (61). He gives an algorithm for the calculation of the v_i in (62). This algorithm sets v_i equal to the integer closest to $q_i m$; if $\Sigma v_i > m$, then the v_i whose fractional parts in excess of 0.5 are smallest are reduced by one until $\Sigma v_i = m$; similarly, if $\Sigma v_i < m$, then the v_i whose fractional parts less than 0.5 are greatest are increased by one until $\Sigma v_i = m$.

Let us denote the v_i that minimize z in (62) subject to (61), by v_i^*. So the number of times that y_i occurs, is fixed now. Nevertheless the <u>order</u> of the occurrences is still undetermined. Sampling is performed in such a way that each order, satisfying $v_i = v_i^*$, has the same probability. Technically this is realized by <u>sampling without replacement</u>. This technique we mentioned in the discussion of

Fig. 2 in Section III.2.4 where we explained Moy's stratification procedure. The technique implies that the probability of sampling the value y_i in drawing t $(t = 1, 2, \ldots, m)$ is

$$(v_i^* - g_i)/(m - t + 1) \tag{63}$$

where g_i denotes the number of times that y_i has already been sampled in this simulation run. Brenner (1963, pp. 295-296) gives a FORTRAN subroutine for calculating v_i^* and for generating a sequence of m values of $\underline{y}$ satisfying $v_i = v_i^*$.

III.3.2. Comment

After having described Brenner's selective sampling technique we shall comment upon it. First of all, we think that Brenner's procedure is biased. For (i) the selective sampling procedure does not realize the theoretical expectation of $\underline{v}_i$, equal to $q_i m$, but realizes $\underline{v}_i$ equal to v_i^*, which is an integer close to $q_i m$ but not necessarily exactly equal to $q_i m$. (ii) Brenner's procedure estimates the expected response given that $\underline{v}_i = v_i^*$ $(i = 1, 2, \ldots, M)$. From Fisz (1967, p. 84) it follows that we can write the expected value of the response $\underline{x}$ as in (64).

$$\begin{aligned} E(\underline{x}) &= \sum_{v} [E(\underline{x} | \underline{v}_1 = v_1, \underline{v}_2 = v_2, \ldots, \underline{v}_M = v_M)] \\ &= \sum \sum \cdots \sum [E(\underline{x} | \underline{v}_1 = v_1, \underline{v}_2 = v_2, \ldots, \underline{v}_M = v_M) \\ &\qquad \times P(\underline{v}_1 = v_1, \underline{v}_2 = v_2, \ldots, \underline{v}_M = v_M)] \end{aligned} \tag{64}$$

where we sum over all possible values of $\underline{v}_1, \underline{v}_2, \ldots, \underline{v}_M$, i.e., over all combinations of the integers $v_1, v_2, \ldots, v_M$ satisfying

$$v_i \geq 0 \qquad \text{and} \qquad \sum_1^M v_i = m$$

Brenner, however, estimates[18)]

$$E(\underline{x}|\underline{v}_1 = v_1^*, \underline{v}_2 = v_2^*, \ldots, \underline{v}_M = v_M^*) \tag{65}$$

In general, the conditional expected value in (65) is not equal to the expected value in (64), even if $v_i^* = E(\underline{v}_i)$. Further we discovered that Brenner's selective sampling amounts to the same thing as the "fixed sequence method" devised by Ehrenfeld and Ben-Tuvia (1962, pp. 258, 263, 274-275). The latter authors admit that their method may give a biased estimator, and this is indeed demonstrated by the example in Table 1 of Ehrenfeld and Ben-Tuvia.

Of course, it is possible that the bias of a sampling procedure is outweighed by the variance reduction so the mean square error, defined earlier in (25) above, is decreased. In the simple queuing systems studied by Ehrenfeld and Ben-Tuvia (1963, pp. 259, 263-264, 274-275) the bias was found to be small indeed and therefore these authors neglected the bias and estimated the variance as if there were no bias. The variance reductions estimated in this way, are 56% in the single-server queuing system and 71% in the two-servers system. In the simple inventory systems investigated by Brenner (1963, p. 293) selective sampling was also found to yield good results. (Brenner did not measure the MSE but he measured the "penalty," i.e., the true inventory cost at the optimum factor combination estimated by simulation minus the cost at the exact optimum determined by the analytical solution.) Neither Brenner not Ehrenfeld and Ben-Tuvia took into account the extra computer time needed by their VRT.

At the beginning of this section we pointed out that Brenner restricts his examples to systems with only one type of input variable which moreover is assumed to be discrete instead of continuous, while further m, the number of input values, is a known constant. Obviously Brenner's procedure can be generalized to comprise systems with more types of input. For in that case, we simply apply (62) and (63) to each type of input. In Ehrenfeld and Ben-Tuvia (1962,

p. 267) we indeed found an example with two types of input. Further, if we have a system with continuous input variables we can form M classes, analogous to (57), and let the class mean stand for y_i in (57). Such a procedure was actually used by Ehrenfeld and Ben-Tuvia in their queuing systems where the interarrival and service times are continuous inputs. These authors formed M classes such that there are 1/M observations in each class, and they used the class median for y_i in our equation (57). Obviously replacing a continuous input variable by a discrete variable creates some bias, besides the bias that is created by replacing (64) by (65), as Ehrenfeld and Ben-Tuvia (1962, pp. 259, 264) admit. The existence of $\underline{m}$, a stochastic number of values of the input variable, is one more serious problem in this VRT. This problem is not mentioned by Brenner or Ben-Tuvia and Ehrenfeld. We could replace m by an estimated value of $\underline{m}$ in (62) and (63). Unfortunately this would disturb the procedure and the v_i^*, the desired number of times y_i occurs given $\underline{m}$ turned out to be m, would not be realized.

Finally, we point out that Orcutt et al. (1961, pp. 33-34, 215, 294-299, 373-374) also applied a kind of selective sampling (without using this term). For they force the number of events of a specific type generated in the simulation experiment, to follow the expected number of events (or to follow the real-world number of events if they are simulating a past period for which they have historical data). Their technique for realizing such a tracking mechanism, unlike that applied by Brenner or Ehrenfeld and Ben-Tuvia does not consist of sampling without replacement. Instead they adjust the original probabilities by a multiplicative factor which measures how much the actual number of events generated until the current period of simulated time, deviates from the expected (or historical) number of events; for details we refer to Orcutt et al. (1961, pp. 296-299). Note that Orcutt et al. (1961, p. 296) suggest that this mechanism does not create bias, but our discussion of (64) and (65) shows why we disagree. Meier et al. (1969, pp. 275-277) briefly discuss some more studies using "tracking devices."

III.4. CONTROL VARIATES OR REGRESSION SAMPLING

III.4.1. Fundamentals of Control Variates

The previous VRT, selective sampling, looked at $\underline{v}_i$, the number of times that the input variable $\underline{y}$ has the value y_i. The control variate technique is a more global instrument since it considers only the average value of the stochastic input variable. (In Section III. 4.2 we shall consider a more general definition of the control variate technique.) The average value of the input $\underline{y}$ in a simulation run where we use $\underline{m}$ values of the input variable, is given by (66).

$$\bar{\underline{y}} = \sum_{t=1}^{\underline{m}} \underline{y}_t/\underline{m} \tag{66}$$

(Note that our notation in (66) with the discrete stochastic variable $\underline{m}$ also covers the case $P(\underline{m} = m) = 1$, i.e., a constant m.) Unlike selective sampling, the control variate technique does not fix $\bar{\underline{y}}$ equal to its theoretical value, say η, where η is defined in (67).

$$\eta = E(\bar{\underline{y}}) \tag{67}$$

So the sampling process itself is not adjusted. Instead, at the end of the run the realized value of $\bar{\underline{y}}$ is compared with its expected or "true" value η. This value η is supposed to be known. For $\underline{y}_t$ in (66) is generated from a known input distribution, e.g., stochastic demand or service time, and therefore indeed we can calculate the expectation of $\underline{y}_t$ which is equal to the expectation of $\bar{\underline{y}}$. If $\bar{\underline{y}}$ differs from its theoretical value η, then this VRT corrects this deviation, using (68).

$$\underline{xc} = \underline{x} - a(\bar{\underline{y}} - \eta) \tag{68}$$

where $\underline{x}$ is the response (e.g., total profit or average waiting time) of one simulation run, $\underline{xc}$ is the estimator in case of a control variable $\underline{y}$, and for the time being a is a constant. Obviously the new estimator $\underline{xc}$ is unbiased. For from (67) and (68) it follows that

$$\begin{aligned} E(\underline{xc}) &= E(\underline{x}) - aE(\bar{\underline{y}} - \eta) \\ &= E(\underline{x}) - a\{E(\bar{\underline{y}}) - \eta\} \\ &= E(\underline{x}) \end{aligned} \tag{69}$$

Its variance is given by (70).

$$\begin{aligned} \operatorname{var}(\underline{xc}) &= \operatorname{var}(\underline{x}) + a^2 \operatorname{var}(\bar{\underline{y}}) - 2a \operatorname{cov}(\underline{x},\bar{\underline{y}}) \\ &= \operatorname{var}(\underline{x}) + a^2 \operatorname{var}(\bar{\underline{y}}) - 2a\rho(x,\bar{y}) \sqrt{\operatorname{var}(\underline{x}) \operatorname{var}(\bar{\underline{y}})} \end{aligned} \tag{70}$$

From (70) we see that the variance of the new estimator is smaller than $\operatorname{var}(\underline{x})$, the variance of the crude estimator, if

(a) for a positive value of the constant a (71a) holds

$$a < 2\rho(\underline{x},\bar{\underline{y}})\sigma(\underline{x})/\sigma(\bar{\underline{y}}) \tag{71a}$$

(b) for a negative value of a (71b) holds

$$a > 2\rho(\underline{x},\bar{\underline{y}})\sigma(\underline{x})/\sigma(\bar{\underline{y}}) \tag{71b}$$

Hence if $\underline{x}$ and $\bar{\underline{y}}$ are positively correlated a variance reduction is realized if the constant a is positive but smaller than the right-hand side of (71a); if they are negatively correlated a should be negative but in absolute value smaller than the absolute value of the right-hand side of (71b). We observe that this <u>sign</u> of the coefficient is intuitively reasonable. For, if $\bar{\underline{y}}$ and $\underline{x}$ are positively correlated then a value of $\bar{\underline{y}}$ higher than its expectation η tends to lead to a value of $\underline{x}$ higher than its expectation, say μ. Further in this example $(\bar{y} - \eta)$ will be positive so $a(\bar{y} - \eta)$ is positive if a is positive. Finally $a(\bar{y} - \eta)$ should be positive since x is higher than its expectation μ in this example so we should subtract from x in (68). The <u>optimal</u> choice of the constant a follows from differentiating (70) and solving for a. This yields the optimal value of a, say, a_0, given in (72).

$$a_0 = \frac{\operatorname{cov}(\underline{x}, \bar{\underline{y}})}{\operatorname{var}(\bar{\underline{y}})}$$

$$= \rho(\underline{x}, \bar{\underline{y}})\ \sigma(\underline{x})/\sigma(\bar{\underline{y}}) \tag{72}$$

This optimal constant, substituted into (70), yields the minimal variance (i.e., the variance of $\underline{xc}_0$, the optimal control variate estimator) given in (73).

$$\operatorname{var}(\underline{xc}_0) = \operatorname{var}(\underline{x})\ \{1 - \rho^2(\underline{x}, \bar{\underline{y}})\} \tag{73}$$

We point out that the limits for the constant a, given in (71a) and (71b), and the optimal choice of a, given in (72), depend on the correlation between the output $\underline{x}$ and the input $\bar{\underline{y}}$. We do not know this correlation and therefore we may decide to estimate it. We shall see in Section III.4.3 how we can estimate the optimal value of the coefficient a and what the consequences are of this estimation. First, however, we shall consider more general definitions of the control variate technique.

III.4.2. Types of Control Variates

In this section we shall describe several types of control variates. The derivation of the optimal coefficients of these new types and the estimation of these coefficients will be presented in the next section.

(i) If there are more types of input variables, e.g., service time and interarrival time, then the following type of "multiple" control variate estimator is obvious.

$$\underline{xcm} = \underline{x} - \sum_{k=1}^{K} a_k(\bar{\underline{y}}_k - \eta_k) \tag{74}$$

with

$$\bar{y}_k = \sum_{t=1}^{m_k} y_{kt}/m_k \tag{75}$$

$$\eta_k = E(\bar{y}_k) \tag{76}$$

where y_{kt} is the t'th observation of the k'th type of input variable in the simulation run. It is easy to see that this estimator is still unbiased. Its variance will depend on the choice of the coefficients a_k, which will be discussed in Section III.4.3.

(ii) Above we generalized the control variate technique by considering more than one type of input variable. A second generalization consists of defining the control variate y of (68) not as an average as in (66) but as some function of the individual input variables y_t. So redefine the control variable, now denoted by z, as in (77).

$$z = g(y_1, y_2, \ldots, y_m) \tag{77}$$

Hence (77) specifies the control variate z as some function g of the individual observations of the stochastic input y_t; we study the general case where m, the number of individual observations, may be stochastic. A necessary condition to be imposed on the function g is that we are able to calculate the expected value of the new control variate. Denote this expected value by ξ, i.e.,

$$E(z) = E[g(y_1, y_2, \ldots, y_m)] = \xi \tag{78}$$

Analogous to (68), (69), and (70) the generalized control variate estimator xcg and its mean and variance are given by (79), (80), and (81), respectively.

$$\begin{aligned} xcg &= x - z + E(z) \\ &= x - g(y_1, y_2, \ldots, y_m) + \xi \end{aligned} \tag{79}$$

with

$$E(\underline{x}cg) = E(\underline{x}) - E(\underline{z}) + E(\xi)$$
$$= E(\underline{x}) - \xi + \xi = E(\underline{x}) \tag{80}$$

and

$$\text{var}(\underline{x}cg) = \text{var}(\underline{x}) + \text{var}(\underline{z}) - 2\,\text{cov}(\underline{x},\underline{z}) \tag{81}$$

From (81) a second condition on the function g follows. For g should be so chosen that the control variate $\underline{z}$ has a "strong" positive correlation with the response $\underline{x}$. This condition replaces the conditions given in (71a) and (71b) for the more specific control variate (68). Moy (1965, pp. 60-61; 1966, pp. 9-10) experimented successfully with

$$\underline{z}_1 = g_1(\underline{y}_1, \underline{y}_2, \ldots, \underline{y}_m) = b_1 + b_2 \sum_{t=1}^{m} \underline{y}_t \tag{82}$$

which after substitution into (79) yields the control variate estimator, say $\underline{x}cl$, given in (83).

$$\underline{x}cl = \underline{x} - (b_1 + b_2 \sum_1^m \underline{y}_t) + E(b_1 + b_2 \sum_1^m \underline{y}_t)$$
$$= \underline{x} - b_2 \left\{ \sum_1^m \underline{y}_t - E(\sum_1^m \underline{y}_t) \right\} \tag{83}$$

This estimator resembles the original simple control variate estimator given in (68). Moy also used the higher-degree expression (84).

$$\underline{z}_2 = g_2(\underline{y}_1, \underline{y}_2, \ldots, \underline{y}_m) = b_1 + b_2 \sum_1^m \underline{y}_t + b_3 (\sum_1^m \underline{y}_t)^2 \tag{84}$$

As we have just remarked, in order to use (84) as a control variate we must be able to calculate its expectation. Let us consider a more general definition of this control variable, namely, the control variable with stochastic $\underline{m}$ (not discussed by Moy). Then the control variate estimator based on (84) but with stochastic $\underline{m}$, is given by (85).

$$\underline{x}c2 = \underline{x} - \left\{ b_1 + b_2 \sum_1^{\underline{m}} \underline{y}_t + b_3 \left(\sum_1^{\underline{m}} \underline{y}_t \right)^2 \right\}$$

$$+ E\left[b_1 + b_2 \sum_1^{\underline{m}} \underline{y}_t + b_3 \left(\sum_1^{\underline{m}} \underline{y}_t \right)^2 \right]$$

$$= \underline{x} - b_2 \left\{ \sum_1^{\underline{m}} \underline{y}_t - E\left(\sum_1^{\underline{m}} \underline{y}_t \right) \right\}$$

$$- b_3 \left\{ \left(\sum_1^{\underline{m}} \underline{y}_t \right)^2 - E\left[\left(\sum_1^{\underline{m}} \underline{y}_t \right)^2 \right] \right\} \tag{85}$$

In Appendix III.7 we derive

$$E\left[\sum_1^{\underline{m}} \underline{y}_t \right] = \eta_1 \, E(\underline{m}) \tag{86}$$

and

$$E\left[\left(\sum_1^{\underline{m}} \underline{y}_t \right)^2 \right] = \eta_1^2 \, E(\underline{m}^2) + \sigma_1^2 \, E(\underline{m}) \tag{87}$$

where the moments $\eta_1 = E(\underline{y}_t)$ and $\sigma_1^2 = var(\underline{y}_t)$ can be calculated from the distribution of the input $\underline{y}_t$. If we can also calculate the values of $E(\underline{m})$ and $E(\underline{m}^2)$, then we can substitute these values into (86) and (87) and use the result in (85). Otherwise we can apply the following control variate estimator.

$$\underline{x}c3 = \underline{x} - b_2 \left\{ \sum_1^{\underline{m}} \underline{y}_t - \eta_1 \underline{m} \right\} - b_3 \left\{ \left(\sum_1^{\underline{m}} \underline{y}_t \right)^2 - (\eta_1^2 \underline{m}^2 + \sigma_1^2 \underline{m}) \right\} \tag{88}$$

It is easy to check, using (86) and (87), that (88) defines an <u>unbiased</u> estimator. So we can generate a simulation run, keeping track of the generated input values y_t; at the end of the run we know which value of $\underline{m}$ and $\Sigma \underline{y}_t$ resulted and we substitute these values into (88).

(iii) We can combine the two generalizations, viz., multiple control variates and a general function g. Moy (1965, p. 60), e.g., experimented with the following function where m_1 and m_2 denote the number of values of input of type 1 and 2 respectively.

$$g_3(\underline{y}_{11}, \underline{y}_{12}, \ldots, \underline{y}_{1m_1}, \underline{y}_{21}, \underline{y}_{22}, \ldots, y_{2m_2})$$

$$= b_1 + b_2 \sum_1^{m_1} \underline{y}_{1t} + b_3\left(\sum_1^{m_1} \underline{y}_{1t}\right)^2 + b_4 \sum_1^{m_2} \underline{y}_{2t} + b_5\left(\sum_1^{m_2} \underline{y}_{2t}\right)^2 \qquad (89)$$

(iv) Since the stochastic input variables are generated from random numbers $\underline{r}$ we can directly define the control variate as a function of the random numbers. Hence Moy (1965, p. 60) and (1966, pp. 10-11) experimented with

$$h_1(\underline{r}_1, \underline{r}_2, \ldots, \underline{r}_m)$$

$$= b_1 + b_2 \sum_1^{m_1} \underline{r}_t + b_3\left(\sum_1^{m_1} \underline{r}_t\right)^2 + b_4 \sum_{m_1+1}^{m} \underline{r}_t + b_4\left(\sum_{m_1+1}^{m} \underline{r}_t\right)^2 \qquad (90)$$

where the first m_1 random numbers are used for generating input of type 1 (e.g., interarrival times) and the remaining random numbers for generating input of type 2 (e.g., service times).

(v) All the above control variates are called system-dependent by Moy. For in (68), (74), (79) and (89) the stochastic input $\underline{y}$ is a transformation of the random number $\underline{r}$ and this transformation depends on the system being simulated. Even (90) is system-dependent since the random numbers are partitioned into two parts because the particular system has two types of inputs. An example of a system-independent control variate is (91).

$$h_2(\underline{r}_1, \underline{r}_2, \ldots, \underline{r}_m) = b_1 + b_2 \sum_1^m \underline{r}_t + b_3\Big(\sum_1^m \quad [illegible] \quad 1)$$

In the discussion of (81) we pointed out that the control variable should be defined in such a way that it has a strong (positive) correlation with the response variable. If we pool all random numbers as in (91) then as Moy (1965, p. 36) and (1966, pp. 5, 9) suggests, we should write the computer program in such a way that increasing random numbers tend to increase the response variable. How this relation can be realized was explained in the discussion of (43) and (44) in Section III.2.4. We point out that in the system-dependent control variates there is no such pooling since each type of input variable is taken separately. The sign of the coefficients b will make the correlation between the control variable and the response positive; see also Section III.4.3.

Moy (1965, pp. 35-36) presents several more types of control variates, but they were found to imply too much extra computer time for their calculation or to yield too weak a correlation with the response variable. The experimental results obtained by Moy and also by Radema (1969, pp. 22-24, 31) further indicate that better results are achieved the more system-dependent the control variate is. This is intuitively acceptable since a system-dependent control variate incorporates more prior knowledge about the system. So it seems better to define the control variable in terms of the input variables than to define the control variable as a function of the random numbers themselves, and further it is wise to keep the several types of inputs separated. The experimental results of Moy (1965, p. 62) and Radema (1969, pp. 23, 47) indicate that the correlation between the control variate and the response is not strengthened by adding higher-degree terms as in (84), (89), (90), and (91). So it is better to put the second order coefficients equal to zero. Even then there may be many coefficients b if there are many types of input in the system. As Moy (1965, p. 63) remarks, we can put the coefficient b_k equal to zero if our prior knowledge about the system suggests that type k of the input variables is not strongly correlated with the output. The fact that there are only a few coefficients b is important since we need to estimate the optimal value of each coefficient. (We shall return to this problem in Section III.4.3.)

(vi) Moy's system-independent but also his system-dependent control variates all have such a form that they can be applied in all kinds of simulation studies. Of course we can form a special control variate for the special type of system we want to simulate. Radema (1969) used

$$\underline{x}_{cr} = \underline{x} - a\{\bar{\underline{w}}/\bar{\underline{v}} - E(\bar{\underline{w}}/\bar{\underline{v}})\} \qquad (92)$$

where $\bar{\underline{v}}$ and $\bar{\underline{w}}$ are the average interarrival time and service time respectively in the single-server queuing system and therefore $\bar{\underline{w}}/\bar{\underline{v}}$ estimates the utilization of the system. For m independent exponentially distributed arrival times $\underline{v}$ and service times $\underline{w}$, Radema (1969, p. 23) proved that

$$E\left(\frac{\bar{\underline{w}}}{\bar{\underline{v}}}\right) = \frac{m}{m-1}\,\frac{E(\bar{\underline{w}})}{E(\bar{\underline{v}})} \qquad (93)$$

Beja (1969, pp. 5-7) experimented with several other control variates in queuing systems with some customers having priority, e.g.,

$$\underline{z} = \sum_{j=1}^{m} \underline{s}_j/m \qquad (94)$$

with

$$\underline{s}_j = \max(\underline{w}_{j-1} - \underline{v}_j,\ 0) \qquad (95)$$

where $\underline{v}_j$ is the time between the arrival of the $(j - 1)$'th and the j'th customer, and $\underline{w}_{j-1}$ is the service time of the $(j - 1)$'th customer. Variables like $\underline{z}$ can be defined per priority class. For the derivation of $E(\underline{z})$ we refer to Beja (1969, pp. 5-6). This author mentions some more control variates that are specific for queuing systems with priority classes. Gaver (1969) discusses other control variates for queuing systems. Raj (1968, pp. 85-106, 140-142, 150-151) investigates control variates not in Monte Carlo experiments but in sample surveys.

(vii) Related to control variates is the so-called ratio estimator, $(\underline{x}/\underline{y})\ E(\underline{y})$. This estimator is biased if $E(\underline{x}/\underline{y}) \neq E(\underline{x})/E(\underline{y})$. In general the estimator is useful if $\underline{x}$ and $\underline{y}$ have a correlation coefficient exceeding 1/2 (Gray and Schucany, 1972, p. 83). To reduce the bias jackknifing may be applied; see the discussion of the jackknife in Appendix 1. For a discussion of ratio estimators, their relation to control variates, and their jackknifing we refer to Rao (1969); see also Gray and Schucany (1972, pp. 80-102), Mosteller and Tukey (1968, pp. 144-147), and Raj (1968, pp. 85-106).

We observe that all control variate types leave the sampling process undistorted. Hence the generated timepath can also be used for the investigation of other aspects than the mean response, especially the dynamic behavior of the system.

III.4.3. Estimation of Control Coefficients

In (71a) and (71b) we saw that the control variate estimator yields a lower variance than the crude estimator only if the control coefficient lies within certain limits. These limits involve the covariance (or the correlation) between the response and the control variable, and this covariance is unknown. The optimum control coefficient is also a function of the unknown covariance as (72) showed. In this section we shall discuss how the optimum control coefficients can be estimated for the several variants of control variates we presented in the previous section.

(i) The optimal coefficient of the simple control variate estimator (68) was derived to be

$$a_0 = \mathrm{cov}(\underline{x},\bar{\underline{y}})/\mathrm{var}(\bar{\underline{y}}) = \rho(\underline{x},\bar{\underline{y}})\sigma(\underline{x})/\sigma(\bar{\underline{y}}) \qquad (96)$$

Now it is possible to calculate $\mathrm{var}(\bar{\underline{y}})$ from (97).

$$\mathrm{var}(\bar{\underline{y}}) = \mathrm{var}(\underline{y}_t)\ E(\underline{m}^{-1}) \qquad (97)$$

It is simple to prove that (97) holds if $\bar{\underline{y}}$ is the average of $\underline{m}$ independent observations $\underline{y}_t$, each $\underline{y}_t$ having the same distribution. The first factor in (97), i.e., $\mathrm{var}(\underline{y}_t)$, can be calculated from the distribution of $\underline{y}_t$. The second factor is simply $1/m$ if we have a known constant m. Otherwise we may approximate $\mathrm{var}(\bar{\underline{y}})$ using (98) where we still have to determine $E(\underline{m})$.

$$E(\underline{m}^{-1}) \approx 1/E(\underline{m}) \tag{98}$$

A more serious problem is that we do not know $\mathrm{cov}(\underline{x},\bar{\underline{y}})$ in (96). As Moy (1965, pp. 32-33) remarks, we may either estimate only $\mathrm{cov}(\underline{x},\bar{\underline{y}})$ or both $\mathrm{cov}(\underline{x},\bar{\underline{y}})$ and $\mathrm{var}(\bar{\underline{y}})$. In one experiment with a single-server system Radema (1969, Table 3.4.1.2) applied both procedures and found the same variance reduction for the two resulting control variate estimators. Estimating both $\mathrm{cov}(\underline{x},\bar{\underline{y}})$ and $\mathrm{var}(\bar{\underline{y}})$ is desirable if an error in one estimate is compensated by an error in the other estimate.[19)] We shall follow the other authors on control variates like Moy, Radema, and Tocher, who estimate both $\mathrm{cov}(\underline{x},\bar{\underline{y}})$ and $\mathrm{var}(\bar{\underline{y}})$. For this procedure may result in a compensation of estimation errors as we just remarked. Moreover we shall see that this procedure yields the familiar least squares algorithm. So let a_0 be estimated by (99).

$$\hat{\underline{a}}_0 = \frac{\underline{c}(\underline{x},\bar{\underline{y}})}{\underline{s}^2(\bar{\underline{y}})} = \underline{r}(\underline{x},\bar{\underline{y}}) \frac{\underline{s}(\underline{x})}{\underline{s}(\bar{\underline{y}})} \tag{99}$$

where

$$\underline{c}(\underline{x},\bar{\underline{y}}) = \sum_{i=1}^{n} (\underline{x}_i - \bar{\underline{x}})(\bar{\underline{y}}_i - \bar{\bar{\underline{y}}})/(n - 1) \tag{100}$$

$$\underline{s}^2(\bar{\underline{y}}) = \sum_{i=1}^{n} (\bar{\underline{y}}_i - \bar{\bar{\underline{y}}})^2/(n - 1) \tag{101}$$

$$\underline{s}^2(\underline{x}) = \sum_{i=1}^{n} (\underline{x}_i - \bar{\underline{x}})^2/(n - 1) \tag{102}$$

$$\underline{r}(\underline{x},\bar{\underline{y}}) = \sum_{i=1}^{n} (\underline{x}_i - \bar{\underline{x}})(\bar{\underline{y}}_i - \bar{\bar{\underline{y}}})/\{(n - 1)\ \underline{s}(\underline{x})\ \underline{s}(\bar{\underline{y}})\} \qquad (103)$$

$$\bar{\bar{\underline{y}}} = \sum_{i=1}^{n} \bar{\underline{y}}_i/n \qquad (104)$$

$$\bar{\underline{x}} = \sum_{i=1}^{n} \underline{x}_i/n \qquad (105)$$

i.e., we use the well-known estimators for the variance, the covariance and the correlation coefficient. (Note that in the literature an individual observation is usually denoted by $\underline{x}_i$ and $\underline{y}_i$ but in our case such an "individual" observation is the response $\underline{x}_i$ of simulation run i and the average input $\bar{\underline{y}}_i$ of run i, where we have n runs.) The curious thing about $\hat{\underline{a}}_0$ is that $\hat{\underline{a}}_0$ is equal to the least squares estimator of the regression coefficient α in the simple linear regression equation given in (106); cf. Johnston (1963, p. 12).

$$\underline{x} = \alpha\,\bar{\underline{y}} + \beta + \underline{u} \qquad (106)$$

where $\underline{u}$ denotes the stochastic disturbance around the regression line. By definition the least squares estimate of α minimizes

$$\sum_{1}^{n} (x_i - \tilde{x}_i)^2 = \sum_{1}^{n} (x_i - \alpha\,\bar{y}_i - \beta)^2 \qquad (107)$$

where $\tilde{x}$ denotes the value predicted from the regression equation.

Estimation of the control coefficient a_0 leads to a complication since $\hat{\underline{a}}_0$ is stochastic and this results in a biased control variate estimator $\underline{xce}$. For the control variate estimator with estimated a_0 is given by (108)

$$\underline{xce}_i = \underline{x}_i - \hat{\underline{a}}_0(\bar{\underline{y}}_i - \eta) \qquad (i = 1, \ldots, n) \qquad (108)$$

Hence

$$E(\underline{x}ce_i) = E(\underline{x}_i) - E(\hat{\underline{a}}_0\, \bar{\underline{y}}_i) + \eta\, E(\hat{\underline{a}}_0) \tag{109}$$

where (110)

$$E(\hat{\underline{a}}_0\, \bar{\underline{y}}_i) = E(\hat{\underline{a}}_0)\, E(\bar{\underline{y}}_i) = E(\hat{\underline{a}}_0)\, \eta \tag{110}$$

would hold only if $\hat{\underline{a}}_0$ and $\bar{\underline{y}}_i$ had been independent, as follows from Fisz (1967, p. 82). However, (99) shows that $\hat{\underline{a}}_0$ depends on $\bar{\underline{y}}_i$. Hence the estimator $\underline{x}ce_i$ is biased and so is the average of the n estimators $\underline{x}ce_i$ $(i = 1, \ldots, n)$. Let us see how this can be remedied.

In order to obtain an unbiased control variate estimator Tocher (1963, p. 115) mentions the possibility of splitting the n pairs $(\underline{x}_i, \bar{\underline{y}}_i)$ into two groups, each group consisting of $n_1 = n/2$ pairs. From group 1 α_1 is to be determined in such a way that (111) is minimized.

$$\sum_{h=1}^{n_1} (x_h - \alpha_1 \bar{y}_h - \beta_1)^2 \tag{111}$$

Denote the resulting least squares estimator of α_1 by $\hat{\underline{a}}_{01}$. From group 2 α_2 is so chosen that (112) is minimized.

$$\sum_{g=n_1+1}^{n} (x_g - \alpha_2 \bar{y}_g - \beta_2)^2 \tag{112}$$

Let $\hat{\underline{a}}_{02}$ denote the least squares estimator of α_2. Combine $\hat{\underline{a}}_{01}$ which is based on group 1, with the observations of group 2! This yields the "control variate estimator with estimated coefficient based on splitting", specified in (113).

$$\underline{x}ces_g = \underline{x}_g - \hat{\underline{a}}_{01}(\bar{\underline{y}}_g - \eta) \qquad (g = n_1+1, \ldots, n) \tag{113}$$

Since $\hat{\underline{a}}_{01}$ is not based on $\bar{\underline{y}}_g$ is is simple to prove that $\underline{x}ces_g$ is unbiased, using (110). Hence from the n pairs $(\underline{x}_i, \bar{\underline{y}}_i)$ we can form the unbiased estimator

$$\bar{\underline{x}}ces = \frac{1}{2}\,\bar{\underline{x}}ces_I + \frac{1}{2}\,\bar{\underline{x}}ces_{II} \tag{114}$$

where

$$\bar{\underline{x}}ces_I = \sum_{g=n_1+1}^{n} \underline{x}ces_g/n_1 \tag{115}$$

and $\bar{\underline{x}}ces_{II}$ is defined analogously by combining $\hat{\underline{a}}_{02}$ with the observations in group 1.

Tocher (1963, p. 116) derived that the variance of $\underline{x}ces$ is twice that of the estimator with non-stochastic a_0, the latter variance being given in (73). This variance increase can be remedied by generalized splitting. Then the n pairs of observations are split into J groups $(2 < J \leq n)$, and $\hat{\underline{a}}_{0j}$ $(j = 1, 2, \ldots, J)$ is based on all pairs except for the pairs in group j. Analogous to (113) we form

$$\underline{x}ceg_{ji} = \underline{x}_{ji} - \hat{\underline{a}}_{0j}(\bar{\underline{y}}_{ji} - \eta) \quad (i = 1, \ldots, n/J) \tag{116}$$

where $\underline{x}_{ji}$ and $\bar{\underline{y}}_{ji}$ form pair i within group j. Tocher derived that the variance of the resulting estimator is lower than that of the estimator based on splitting into only two groups. An obvious disadvantage of the generalized procedure is that more computer time is needed since more coefficients a_{0j} are estimated, especially if J is high. The splitting technique can also be found in the statistical literature on jackknifing regression (and ratio) estimators for bias reduction; see e.g. Rao (1969, p. 216), or, more generally, Gray and Schucany (1972, pp. 80-107) and Raj (1968, p. 236). In the jackknife literature it is recommended to take as many groups as possible, or $J = n$; see Rao (1969, p. 217). In their simulation studies Gaver and Shedler (1971, p. 449) and Radema (1969, p. 25) took $J = n$.[20]

(ii) Next we consider the more general case where we have multiple control variables $\bar{\underline{y}}_k$ $(k = 1, \ldots, K)$ as specified in (74). It will be convenient to have control variables with zero-expectation. Hence we define new control variables $\underline{z}_k$:

$$\underline{z}_k = \bar{\underline{y}}_k - E(\bar{\underline{y}}_k) = \bar{\underline{y}}_k - \eta_k \tag{117}$$

so that

$$E(\underline{z}_k) = 0 \tag{118}$$

Then the control variate estimator (74) is equivalent to (119).

$$\underline{x}\text{cm} = \underline{x} - \sum_1^K a_k \underline{z}_k \tag{119}$$

It is realistic to assume that a control variable $\underline{z}_k$ is <u>independent</u> of another control variable $\underline{z}_{k'}$, $(k \neq k')$. For in most simulation models the input $\underline{y}_k$ is independent of the input $\underline{y}_{k'}$. For instance, the service time at one counter is usually supposed to be independent of the service time at another counter or independent of the inter-arrival time. (Later on in this section we shall consider dependent control variables.) For independent control variables (119) yields

$$\text{var}(\underline{x}\text{cm}) = \text{var}(\underline{x}) + \sum_1^K a_k^2 \text{ var}(\underline{z}_k) - 2 \sum_1^K a_k \text{ cov}(\underline{x}, \underline{z}_k) \tag{120}$$

Hence the optimal coefficients can be determined by partial differentiation of (120) and solving for a_k. This gives the optimal values a_{k0}:

$$a_{k0} = \frac{\text{cov}(\underline{x}, \underline{z}_k)}{\text{var}(\underline{z}_k)} \qquad (k = 1, 2, \ldots, K) \tag{121}$$

Since $E(\underline{z}_k)$ is zero, (121) is equivalent to (122).

$$a_{k0} = \frac{E(\underline{x}\ \underline{z}_k)}{E(\underline{z}_k^2)} \qquad (k = 1, 2, \ldots, K) \tag{122}$$

Hence a simple[21)] estimator of the optimal control coefficient is given by (123).

$$\underline{\hat{a}}_{k0} = \frac{\sum_{i=1}^{n} \underline{x}_i \underline{z}_{ki}}{\sum_{i=1}^{n} \underline{z}_{ki}^2} \qquad (k = 1, 2, \ldots, K) \tag{123}$$

Now consider the <u>linear regression</u> equation in (124)

$$\underline{x}_i = \sum_{k=1}^{K} a_k \underline{z}_{ki} + \underline{u}_i \qquad (i = 1, 2, \ldots, n) \tag{124}$$

or its matrix equivalent

$$\underline{\vec{X}} = \underline{\vec{Z}} \vec{A} + \underline{\vec{U}} \tag{125}$$

with

$$\underline{\vec{X}}' = (\underline{x}_1, \underline{x}_2, \ldots, \underline{x}_n) \tag{126}$$

$$\underline{\vec{Z}} = \begin{bmatrix} \underline{z}_{11}, & \underline{z}_{21}, & \ldots, & \underline{z}_{K1} \\ \underline{z}_{12}, & \underline{z}_{22}, & \ldots, & \underline{z}_{K2} \\ \vdots & \vdots & & \vdots \\ \underline{z}_{1n}, & \underline{z}_{2n}, & \ldots, & \underline{z}_{Kn} \end{bmatrix} \tag{127}$$

$$\vec{A}' = (a_1, a_2, \ldots, a_K) \tag{128}$$

$$\underline{\vec{U}}' = (\underline{u}_1, \underline{u}_2, \ldots, \underline{u}_n) \tag{129}$$

From Johnston (1963, p. 109) it follows that the least squares estimators[22] of $\vec{A}$ in (125) are given by (130).

$$\underline{\hat{\vec{A}}} = (\underline{\vec{Z}}' \underline{\vec{Z}})^{-1} \underline{\vec{Z}}' \underline{\vec{X}} \tag{130}$$

The element in row k and column k' of the matrix $(\vec{\underline{Z}}'\vec{\underline{Z}})$ is given by (131).

$$\sum_{i=1}^{n} \underline{z}_{ki}\, \underline{z}_{k'i} \qquad (k, k' = 1, 2, \ldots, K) \tag{131}$$

Hence this element dividend by $(n - 1)$ is an unbiased estimator of the covariance between $\underline{z}_k$ and $\underline{z}_{k'}$. In the derivation of the optimal control coefficients we assumed that the control variables $\underline{z}_k$ are independent. Therefore we may decide to use this prior knowledge in the estimation of the regression coefficients of (124), i.e., put the off-diagonal elements of $(\vec{\underline{Z}}'\, \vec{\underline{Z}})$ equal to zero. On the main diagonal remain the elements $\Sigma_i\, \underline{z}_{ki}^2$ as (131) shows. Hence $(\vec{\underline{Z}}'\, \vec{\underline{Z}})^{-1}$ reduces to a diagonal matrix with on its diagonal the elements $(\Sigma_i\, \underline{z}_{ki}^2)^{-1}$. Summarizing, the <u>least squares</u> estimators of the regression coefficients in (124) adjusted for our prior knowledge of the independence between the variables $\underline{z}_k$, are identical to the <u>estimated optimal</u> control coefficients (123).

(iii) Next we shall consider the <u>genralized</u> control variable $\underline{z}$ in (132).

$$\underline{z} = a_1 + a_2 g_2(\underline{r}_1, \underline{r}_2, \ldots, \underline{r}_{\underline{m}}) + a_3 g_3(\underline{r}_1, \underline{r}_2, \ldots, \underline{r}_{\underline{m}}) + \cdots + a_K g_K(\underline{r}_1, \underline{r}_2, \ldots, \underline{r}_{\underline{m}}) \tag{132}$$

where g_k denotes a function of the $\underline{m}$ random numbers $\underline{r}_i$ $(i = 1, 2, \ldots, \underline{m})$. So (132) comprises the various types of control variables presented in Section III.4.2. As we saw in that section a condition to be imposed upon g_k is that its expectation can be calculated. Hence we can also consider the control variables $\underline{z}_k$ each having expectation zero, defined in (133).

$$\underline{z}_k = g_k(\underline{r}_1, \underline{r}_2, \ldots, \underline{r}_{\underline{m}}) - E[g_k(\underline{r}_1, \ldots, \underline{r}_{\underline{m}})] \qquad (k = 1, 2, \ldots, K) \tag{133}$$

Note that from (132) and (133) it follows that

$$\underline{z}_1 = 1 - 1 = 0 \tag{134}$$

So the control variate estimator becomes

$$\underline{xcgz} = \underline{x} - \sum_{1}^{K} a_k \underline{z}_k \tag{135}$$

The right-hand side of (135) is identical to that of (119). In (135), however, it is very well possible that the control variables $\underline{z}_k$ are dependent. This dependence indeed exists in (88) through (91) where the second-order terms are dependent on the corresponding first-order terms. Beja (1969, p. 2) derived that the optimal values of the coefficients a_k in (135) are given by (136).[23)]

$$\vec{A}_0 = \vec{\Omega}_z^{-1} \vec{\Psi} \tag{136}$$

where $\vec{\Omega}_z$ is the covariance matrix of the control variates $\underline{z}_k$, i.e.,

$$\begin{aligned} \vec{\Omega}_z &= \{E[(\underline{z}_k - E(\underline{z}_k))(\underline{z}_{k'} - E(\underline{z}_{k'}))]\} \\ &= \{E(\underline{z}_k \underline{z}_{k'})\} \qquad (k, k' = 1, 2, \ldots, K) \end{aligned} \tag{137}$$

the last equality following from (133), and $\vec{\Psi}$ in (136) denotes the vector of covariances between $\underline{x}$ and $\underline{z}_k$, i.e.,

$$\vec{\Psi} = \{E(\underline{x}\ \underline{z}_1), E(\underline{x}\ \underline{z}_2), \ldots, E(\underline{x}\ \underline{z}_k)\} \tag{138}$$

where we used (139)

$$\begin{aligned} \operatorname{cov}(\underline{x}, \underline{z}_k) &= E(\underline{x}\ \underline{z}_k) - E(\underline{x})\, E(\underline{z}_k) \\ &= E(\underline{x}\ \underline{z}_k) \end{aligned} \tag{139}$$

So we may estimate the optimal coefficients a_{0k} by using the sample analog of (136). This yields

$$\hat{\vec{A}}_0 = (\vec{Z}' \vec{Z})^{-1} \vec{Z}' \vec{X} \tag{140}$$

where $\vec{X}$ and $\vec{Z}$ were defined in (126) and (127). Comparing (140) with (130) shows that the optimal coefficients of the control variates in (135) can be calculated by applying the algorithm for the calculation of the least squares coefficients in the linear regression equation (125).[24] For that reason the control variate technique with estimated coefficients is often called "regression sampling." Analogous to (108) through (110) we see that (135) would yield a biased estimator since the coefficients have become dependent on the control variates. But as in (116) this can be remedied by splitting the n observations on x, z_1, ..., z_K into J groups $(2 \leq J \leq n)$ of n/J observations each and estimating a_{kj}, the coefficient for control variable k in group j $(j = 1, ..., J)$, from the other $(n - n/J)$ observations. This yields the "generalized control variate estimator with estimated optimal coefficients" defined in (141).

$$xcze_{ji} = x_{ji} - \sum_{k=1}^{K} \hat{a}_{kj} z_{kji}$$

$$(j = 1, ..., J) \; (i = 1, ..., n/J) \tag{141}$$

Even while we do not know $E(\hat{a}_{kj})$ we do know that $E(\hat{a}_{kj} z_{kji}) = 0$ since $\hat{a}_{kj}$ is made independent of z_{kji} and $E(z_{kji}) = 0$.

We point out that the estimation procedure for the coefficients requires that there are "enough" observations, i.e., replicated simulation runs. As it is observed in Kleijnen (1969, pp. 291-292) the number of runs may be small in simulation studies of steady-state behavior. Such a small number of runs can make it impossible to estimate many control coefficients. Even if it is possible to estimate these coefficients the variances of the estimated coefficients increase as the number of observations decreases, and this leads to less variance reduction. These considerations suggest that a control variate estimator be used with only a few coefficients, unless we have very many observations. So a first-degree control variate as in (82) is more attractive from this point of view than a second degree control variate as in (88) through (91). We also refer to the numerical results of Moy (1965, pp. 49-54). Radema (1969, pp. 23, 47) even found a variance increase when using a quadratic control variate as in (89) instead of using the corresponding first degree control variate.

A problem not mentioned in the simulation literature on control variates is the <u>estimation of the variance</u> of the control variate estimator given in (141). In Appendix III.7 we prove that the control variate estimators $\underline{xcze}_{ji}$ do not show correlation <u>within</u> the same group but do show correlation <u>between</u> groups. The last result is quite obvious since the estimator $\underline{xcze}_{ji}$ depends on $\hat{\underline{a}}_{kj}$ but $\hat{\underline{a}}_{kj}$ depends on the variates outside group j. But if the estimators $\underline{x}cze$ within group j depend on the estimators $\underline{x}cze$ outside group j via $\hat{\underline{a}}_{kj}$, then (142) holds only <u>approximately</u>.

$$v\hat{\underline{a}}r(\bar{\bar{\underline{x}}}cze) = \sum_{j=1}^{J} v\hat{\underline{a}}r(\bar{\underline{x}}cze_j)/J \qquad (142)$$

since $\bar{\underline{x}}cze_j$, the average of group j, is not independent of the averages of the other groups. We further observe that Beja (1969, p. 7) gives an incorrect formula for the variance of the control variate estimator since he uses a formula like (120) replacing $var(\underline{x})$, $var(\underline{z}_k)$ and $cov(\underline{x}, \underline{z}_k)$ by their estimators. A formula like (120), however, is valid only for nonstochastic control coefficients a_k. Radema (1969, pp. 39-40) derived an expression for the variance of an individual control variate estimator, taking into account the stochastic character of $\hat{\underline{a}}_k$. Yet the variance of the average of the individual estimators again involves the unknown covariances between the estimators. Obviously, because of the above dependence a <u>confidence interval</u> like (12) will hold only approximately.

The following approaches may be based on the general literature on sampling techniques.

Rao (1969, p. 222) found that in the cases he studied, the biased estimator (108) (or t_8 in Rao's notation) where a is estimated from all observations, has smaller MSE than the unbiased estimator (116) (or t_9 in Rao) based on generalized splitting (with $J = n$ groups). Therefore we might decide to use the biased estimator, and estimate its variance from the asymptotic relation derived in Raj (1968, p. 100):

$$\text{var}(\bar{\underline{x}}ce) = \sigma_x^2(1 - \rho^2)/n \tag{143}$$

Obviously the resulting confidence interval is based on crude approximations (large sample variance, neglected bias; Rao (1969, p. 224) found that his related variance estimator v_3 has considerable bias). More research is required to find a good estimator for the variance of the regression estimator, this estimator possibly being based on splitting or jackknifing; see also Mickey (1959, pp. 600-601) and Rao (1969, pp. 216-217). One more alternative we would propose, is the following <u>jackknifed</u> regression estimator.

(1) Calculate the (biased) estimator of $\theta = E(\underline{x})$.

$$\underline{\hat{\theta}} = \bar{\underline{x}} - \sum_{k=1}^{K} \underline{\hat{a}}_k \bar{\underline{z}}_k \tag{144}$$

where $\bar{\underline{x}}$ is the average of the responses of n replicated runs, $\bar{\underline{z}}_k$ is the average of the n values of the control variate $\underline{z}_k$, and $\underline{\hat{a}}_k$ is the estimator of the optimal coefficient estimated from the <u>same</u> n runs; compare (108).

(2) Form N groups, each group consisting of $M = n/N$ observations, and calculate $\underline{\hat{\theta}}_{-j}$ analogous to (144) after deleting group j:

$$\underline{\hat{\theta}}_j = \underline{\bar{x}}_j - \sum_{k=1}^{K} \underline{\hat{a}}_{kj} \, \underline{\bar{z}}_{kj} \qquad (j = 1, \ldots, N) \tag{145}$$

where $\underline{\bar{x}}_j$ is the average of the $n - M$ observations remaining after deleting group j; $\underline{\hat{a}}_{kj}$ is estimated from the <u>same</u> $n - M$ observations, etc. Note that contrary to (141) the estimator $\underline{\hat{\theta}}_j$ remains biased. The "pseudovalues" of the jackknife statistic are:

$$\underline{J}_j = N\underline{\hat{\theta}} - (N - 1)\underline{\hat{\theta}}_j \tag{146}$$

(3) Calculate the jackknife estimator:

$$\underline{\bar{J}} = N\underline{\hat{\theta}} - (N - 1) \sum_{j=1}^{N} \underline{\hat{\theta}}_j / N$$

$$= \underline{\bar{x}} + (N - 1) \sum_j \sum_k \underline{a}_{kj} \, \underline{\bar{z}}_{kj} / N \tag{147}$$

Under mild conditions this estimator will have lower bias than $\underline{\hat{\theta}}$ in (144); see the general exposé on the jackknife in Appendix 1.

(4) Calculate a confidence interval from $\underline{\bar{J}}$ treating the $\underline{J}_j$ as independent observations and using the t-distribution. A conservative confidence interval may be constructed by estimating a possible positive correlation among the $\underline{J}_j$; see Appendix 1.

III.4.4. <u>Applications of Control Variates in Simulation and Monte Carlo Studies</u>

Moy (1965, 1966) applied various control variates to single-server queuing systems and to the truck dock system mentioned before. Radema (1969) also applied control variates to a single-server queuing system. In part (v) of Section III.4.2 we observed that their experiments indicate that better results are obtained the more <u>system-dependent</u> the control variate is. In Section III.4.3 we further remarked that we should take a control variate estimator with only a <u>few coefficients</u>. Obviously we can obtain better estimates of these coefficients as we have more observations, i.e., replicated runs available; experimental results can be found in Radema (1969, Table 3.4.1.2).

Both the experimental results of Moy (1965, pp. 55, 60-62, 1966, p. 29) and Radema (1969, Tables 3.4.1.1 and 3.4.1.2) show that worthwhile variance reductions can indeed be obtained with control variates in simple queuing systems. The reductions become more important as the utilization of the system increases. Moy found reductions between 20 and 50%; Radema between 15% (for a utilization factor of only 50%, 10 replicated runs and a control variate estimator with two coefficients as (74) gives for $K = 2$) and 68% (for a utilization of 90%, 40 runs, one coefficient as in (92) above). Moy further applied control variates to the complex truck dock system. The system-independent control variate specified in (91) yielded a variance reduction of only about 7% for 90 replicated runs (Moy, 1965, pp. 56-57). From Moy (1966, p. 33) it follows that in the truck dock system, system-dependent variants gave variance-reductions ranging from 14% to 47%. Beja (1969, p. 6) applied a control variate based on (94) which yielded a variance reduction of 60%, and he applied (94) combined with the average service time and the average arrival time as control variates, and this resulted in a reduction of 92%. Beja (1969, p. 8) further reports that with priority queuing systems "in a number of exploratory trials with no more than four control variates, the variance ratio was over 2 under unfavorable circumstances and over 10 under favorable ones." Andréasson (1971) gives detailed results on the use of control variates in the simulation of a simple telephone exchange, where a certain probability p is to be estimated. For some system variants he halved the variance, but the variance reduction decreased as p approached 1 or zero. Obviously, the above variance-reductions should be corrected for the extra computer time needed for the calculation of the values of the control variates and their optimal coefficients. Moy (1966, p. 33) also calculated these corrected variance reductions for the truck dock system. These reductions range from 1 to 41%. As Radema (1969,

p. 32) remarked the extra computer time becomes less important the more computer time it takes to obtain one simulation run. We might be tempted to save computer time by not estimating the optimal control coefficients but instead using values from previous simulation experiments with related systems. Unfortunately, Moy (1965, pp. 67-70) found that the estimated optimal control coefficients vary much with the system variant, and therefore it seems necessary to estimate the coefficients for each variant separately. Radema (1969) investigated how the variance-reduction is influenced by the utilization factor of the single-server queuing system, the length of an individual simulation run, the starting conditions and the number of replicated runs. Further Gaver (1969) and Lombaers (1968, p. 253) experimented with a control variate in the simple queuing system and Tocher (1963, pp. 115-117, 172-173) also briefly discusses control variates in simulation.

The application of the control variate technique in Monte Carlo studies is straightforward. For instance, in the example of Appendix I.1 we estimated the value of the integral

$$\xi = \int_{v}^{\infty} \frac{1}{y} \lambda e^{-\lambda y} \, dy \qquad (\lambda, v > 0) \tag{148}$$

by sampling $\underline{y}$ from the exponential density function

$$f(y) = \lambda e^{-\lambda y} \qquad (y > 0) \tag{149}$$

Therefore an obvious control variate is $\underline{y}$ with the known mean $1/\lambda$. In Appendix I.2 we considered an example of distribution sampling where we wanted to estimate p, the probability in (150).

$$p = P(\underline{x} < a) \tag{150}$$

where

$$\underline{x} = \min(\underline{y}_1, \underline{y}_2) \tag{151}$$

$\underline{y}_i$ having the normal distribution $N(\eta_i, \sigma_i^2)$ $(i = 1, 2)$. So obvious control variables are $\underline{y}_i$ with the known means η_i. (Actually we did not apply the control variate technique to these two Monte Carlo examples so we cannot give numerical results for the efficiency gain.)

Another kind of control variate can be found in the literature.[25)] Here the control variate is not the stochastic input of the simulated system. Instead a related simpler system is simulated together with the system of interest. This related system is assumed to have a known mean, say μ_z. It is simulated using the same sequence of random numbers as the system of interest. So $\underline{z}$, the simulation response of the simple system, is correlated with $\underline{x}$, the simulation response of the system of interest; see Section III.7. We can estimate μ_x, the mean of the system of interest, from (152).[26)]

$$\underline{\hat{\mu}}_x = \underline{x} - a(\underline{z} - \mu_z) \tag{152}$$

where an optimal value of a can again be estimated if we have replicated runs of both systems. This approach is related to the type of stratification in simulation presented at the beginning of Section III.2.4 and we have the same objections against (152) as we have against that stratification. First, there must be a related system with a known mean. Second, it may take much time to program and run this related system. (It may be, however, that such a simplified related system has already been programmed and run for purposes of validation, verification and sensitivity testing; see Chapter II.) In distribution sampling the extra time for generating the related response $\underline{z}$ may very well be negligible. For consider the estimation of the expected range of a sample of m observations from a normal population. From the m observations we can estimate not only the range, but also the variance. So $\underline{x}$ in (152) stands for the range and $\underline{z}$ for the variance. This example is also given by Tocher (1963, pp. 102-103). Another distribution sampling example is given by Hillier and Lieberman (1968, pp. 460-461) where in (152) $\underline{x}$ is some complicated function of a random number $\underline{r}$, while $\underline{z}$ is a simple function of $\underline{r}$ with a known mean. In these distribution sampling examples the extra programming and computer time for the determination of $\underline{z}$ in (152) may be negligible. Burt et al. (1970) succeeded in (roughly) halving the variance using a control variate like (152) in their stochastic-network simulation. They do not report the extra computer time needed for simulating a simple network with known solution. They further mention the use of more than one control system. Maxwell (1965, pp. 7-9) discusses the use of (152) with $\underline{z}$ being either the stochastic input of the system of interest or the stochastic output of a related system. He also discusses $\underline{z}$ being the response of a related system with unknown mean where μ_z in (152) is now replaced by $\hat{\underline{\mu}}_z$, the estimated response of this system obtained in simulation runs performed in the past. Obviously in the latter case it is harder to realize a variance reduction since $\hat{\underline{\mu}}_z$ increases the variance of the control variate estimator.[27)] Lewis (1972, p. 11) proposes estimating one system variant from a very large number of runs, and using the resulting estimator as a control variate when estimating the other system variants in fewer runs (based on the same random number stream). Or

$$\hat{\underline{\mu}}_x = \bar{\underline{x}}_m - a(\bar{\underline{z}}_m - \bar{\underline{z}}_n) \tag{153}$$

where $\bar{\underline{x}}_m$ and $\bar{\underline{z}}_m$ are the average response of system variant 1 and 2 using the same random numbers, and $\bar{\underline{z}}_n$ is the average of system 2 based on n runs which include the m ($\ll n$) runs of $\bar{\underline{z}}_m$.

Finally, in the literature a procedure can be found that is also called a control variate technique. Nevertheless it differs considerably from the above control variates. For instead of estimating only the mean response it assumes that the complete distribution of the response is to be estimated. This procedure is again based on the assumption that besides the response of interest we observe another variable with a known distribution. The technique was first presented by Fieller and Hartley (1954) and is also described by Farmer (1968, pp. 7-9) and Tocher (1963, pp. 90-93). The latter two authors illustrate the method with a Monte Carlo example where the distribution of the median is to be estimated while the mean is also observed and is supposed to have a known distribution. Matérn (1962) describes several variants of this control variate distribution estimator.

III.5. IMPORTANCE SAMPLING

III.5.1. Fundamentals of Importance Sampling

An elementary discussion of importance sampling can be found in Clark (1961). The basic idea of importance sampling is to disturb the original sampling process completely replacing the original process by another one. This distortion is corrected by some kind of weighing of the observations from the new sampling process, so that the average of the corrected observations is still an unbiased estimator of the mean of the original process. Hence unlike control variates, importance sampling disturbs the original sampling process and consequently the output of the new sampling process cannot be used to investigate the dynamic behavior of the system of interest. Importance sampling looks more like stratified sampling, for (i) the original sampling process is changed (though importance sampling uses a much more distorted process than stratification), (ii) the observations have different weights. We refer to Clark (1961, pp. 608-610), Handscomb (1968, pp. 6-7) and Hillier and Lieberman (1968, p. 458) for a further discussion of the similarities between stratified and importance sampling. Next we shall give a more exact presentation of the fundamental idea of importance sampling as it can also be found in Kahn and Marshall (1953, pp. 269-270).

Suppose we want to estimate ξ, the value of the integral in (154).

$$\xi = \int_{-\infty}^{\infty} g(x)\, f(x)\, dx \qquad (154)$$

where $f(x)$ is a density function. Hence ξ is the expected value of $g(\underline{x})$, if $\underline{x}$ has the density function $f(x)$. Or

$$\xi = E[g(\underline{x})] \tag{155}$$

So a crude estimation procedure would consist of sampling n values of $\underline{x}$ from $f(x)$ and calculating

$$\hat{\underline{\xi}} = \frac{1}{n} \sum_{i=1}^{n} g(\underline{x}_i) \tag{156}$$

An example was given in Appendix I.1 where $f(x)$ and $g(x)$ were specified in (1.3) and (1.4), respectively. However, we can also write (154) as

$$\xi = \int_{-\infty}^{\infty} \frac{g(x)\ f(x)}{h(x)}\ h(x)\ dx \tag{157}$$

So if $h(x)$ is (another) density function of $\underline{x}$ then we can estimate ξ by sampling $\underline{x}$ from $h(x)$ and substituting $\underline{x}$ into (158)

$$g^*(\underline{x}) = \frac{g(\underline{x})\ f(\underline{x})}{h(\underline{x})} = g(\underline{x}) \left[\frac{f(\underline{x})}{h(\underline{x})}\right] \tag{158}$$

where $f(\underline{x})/h(\underline{x})$ can be interpreted as a weighing factor. The expectation of $g^*(\underline{x})$ is given by

$$E[g^*(\underline{x})] = \int_{-\infty}^{\infty} g^*(x)\ h(x)\ dx \tag{159}$$

which reduces to ξ. We can sample $\underline{x}$ from $h(x)$ n times and estimate ξ by

$$\underline{\xi}^* = \frac{1}{n} \sum_{i=1}^{n} g^*(\underline{x}_i) \tag{160}$$

In Kahn and Marshall (1953, p. 270) it is derived that the variance of the importance sampling estimator is minimized if we select as new density function

$$h_0(x) = \frac{|g(x)|\ f(x)}{\int_{-\infty}^{\infty} |g(x)|\ f(x)\ dx} \tag{161}$$

If $g(x) \geq 0$ for all x then (161) reduces to

$$h_0(x) = \frac{g(x)\ f(x)}{\int_{-\infty}^{\infty} g(x)\ f(x)\ dx} = \frac{g(x)\ f(x)}{\xi} \tag{162}$$

So the optimal new density function $h_0(x)$ implies that we sample heavily from the "important" region of $\underline{x}$, i.e., from the region where $\underline{x}$ yields a high value of the response $g(\underline{x})$ (unless the probability of such a value of $\underline{x}$ is small, i.e., $f(x)$ is small). Kahn and Marshall (1953, p. 271) proved that if $g(x) \geq 0$ for all x, then the variance of the optimal importance sampling estimator becomes zero. Unfortunately we cannot calculate the optimal density (162) since (162) contains ξ and ξ is to be estimated! Nevertheless, (162) can be used to approximate the optimal density function $h_0(x)$. We shall first demonstrate how much an approximation can be found in a Monte Carlo study. Next we shall present Moy's approximation for the more complicated case of simulation.

III.5.2. An Example of Importance Sampling in a Monte Carlo Study[28)]

Suppose we want to estimate the integral studied before in Appendix I.1 and given again in (163).

$$\xi(\lambda, v) = \int_{v}^{\infty} \frac{1}{x} \lambda e^{-\lambda x}\ dx \qquad (\lambda,\ v > 0) \tag{163}$$

Comparing (163) with the general expression for ξ in (154) shows that in this problem we have

$$g(x) = 0 \quad \text{if} \quad x < v$$
$$= \frac{1}{x} \quad \text{if} \quad x \geq v \tag{164}$$

and

$$f(x) = \lambda e^{-\lambda x} \quad \text{if} \quad x \geq 0$$
$$= 0 \quad \text{if} \quad x < 0 \tag{165}$$

where (164) and (165) are identical with (1.3) and (1.4) in Appendix I.1. From (162) it follows that the optimal density function for importance sampling in this problem is given by

$$h_0(x) = 0 \quad \text{if} \quad x < v$$
$$= \frac{\lambda}{\xi}\frac{1}{x}e^{-\lambda x} \quad \text{if} \quad x \geq v \tag{166}$$

This optimal density function is pictured in Fig. 3.

We shall discuss three <u>approximations</u> to this optimal distribution. (Let an asterisk * denote an importance sampling estimator and let the suffixes 1, 2 and 3 denote approximation (i), (ii), and (iii), respectively.)

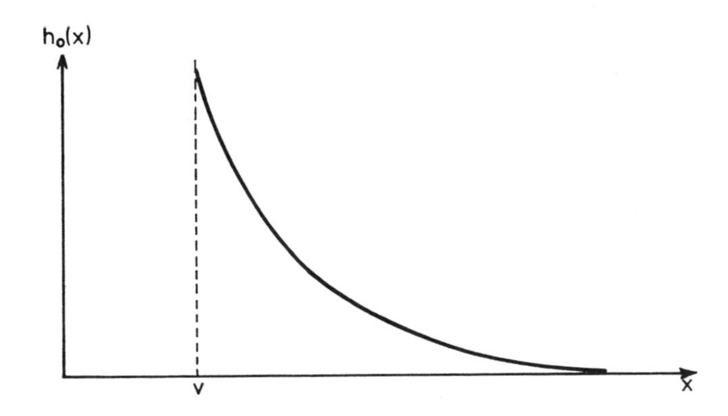

Fig. 3. The optimal importance sampling density function for the estimation of $\xi(\lambda, v)$.

(i) The shifted exponential distribution. We first sample $\underline{x}'$ from the exponential distribution with parameter c (instead of the parameter λ of (165); the value of c is determined later on). Next we increase this $\underline{x}'$ with the constant v (this is the constant of Fig. 3). The density function $h_1(x)$ of the shifted exponential $\underline{x} = \underline{x}' + v$ is given by (167) as can be checked using e.g. Fisz (1967, pp. 39-40).

$$
\begin{aligned}
h_1(x) &= ce^{-c(x-v)} = ce^{cv}\, e^{-cx} && \text{if } x \geq v \\
&= 0 && \text{if } x < v
\end{aligned}
\tag{167}
$$

If we select $h_1(x)$ in (167) as the new density function from which we generate $\underline{x}$ in importance sampling, then the new $g(\underline{x})$ becomes $g_1^*(\underline{x})$ and can be determined from (158), (164), (165) and (167).

$$
\begin{aligned}
g_1^*(x) &= (\tfrac{1}{x})\,(\lambda e^{-\lambda x})\,(ce^{cv}\, e^{-cx})^{-1} \\
&= \frac{\lambda}{c}\, e^{-cv}\, \frac{1}{x}\, e^{-(\lambda-c)x}
\end{aligned}
\tag{168}
$$

where $x \geq v$ since (167) implies that we cannot generate $x < v$. From (160) we see that after having generated n values of $\underline{x}$ from $h_1(x)$, we estimate ξ by

$$
\underline{\xi}_1^* = \frac{1}{n} \sum_{i=1}^{n} g_1^*(\underline{x}_i) \tag{169}
$$

Next we have to determine the value of the constant c in (167). We like to choose a value of c such that the variance of $\underline{\xi}_1^*$ is minimized for a given n, i.e., the variance of $g_1^*(\underline{x})$ is to be minimized. In Kleijnen (1968b, p. 195-197) it is derived that this optimal c cannot be calculated since the expression for the optimal c involves the unknown ξ again. Therefore the following approach is suggested in Kleijnen (1968b, p. 183). In Section III. 5.1 we saw that the optimal density function $h_0(x)$ in (162) implies that the variance of the importance sampling estimator is zero. Consequently its range R is also zero, R being defined in (170).

$$R[g_1^*(\underline{x})] = \max_x[g_1^*(x)] - \min_x[g_1^*(x)] \quad (170)$$

Since we cannot calculate c in such a way that the variance of $g_1^*(\underline{x})$ is minimized, we so select c that the range of $g_1^*(\underline{x})$ is minimized. If $c > \lambda$ then, as (168) shows, $g_1^*(x) = \infty$ for $x = \infty$, or $R = \infty$. So we should take $c \leq \lambda$ in order to keep the range finite. For any value of c[29] in the interval $(0,\lambda]$ the minimal value of $g_1^*(x)$ is 0 if $x = \infty$. So to minimize R we have to minimize the maximum value of $g_1^*(x)$, by selection of an appropriate value of c in $(0,\lambda]$. This $\max[g_1^*(x)]$ is reached for $x = v$ since (168) shows that $g_1^*(x)$ is a decreasing function of x (if $0 < c \leq \lambda$), i.e., (171) holds.

$$\max_x[g_1^*(x)] = g_1^*(v) = \frac{\lambda}{c} e^{-cv} \frac{1}{v} e^{-(\lambda-c)v}$$

$$= \frac{\lambda}{c} \frac{1}{v} e^{-\lambda v} \quad (171)$$

Hence $\max[g_1^*(x)]$ is minimized if we select a value of c as large as possible in the interval $(0,\lambda]$, or the range of R is minimal if $c = \lambda$. Substitution of $c = \lambda$ into (167) yields

$$\begin{aligned} h_1(x) &= \lambda e^{\lambda v} e^{-\lambda x} && \text{if } x \geq v \\ &= 0 && \text{if } x < v \end{aligned} \quad (172)$$

i.e., we sample $\underline{x}'$ from the original density function given in (165) and increase $\underline{x}'$ with the constant v. This new $\underline{x}$ is substituted into (173) which results from (168) if we put $c = \lambda$.

$$g_1^*(\underline{x}) = e^{-\lambda v} \frac{1}{\underline{x}} \quad (\underline{x} > v) \quad (173)$$

We observe that $g(\underline{x})$, the crude Monte Carlo estimator in (164), has a larger range than $g_1^*(\underline{x})$. For

$$R[g_1^*(\underline{x})] = e^{-\lambda v}\frac{1}{v} - e^{-\lambda v}\frac{1}{\infty} = e^{-\lambda v}\frac{1}{v} \tag{174}$$

and

$$R[g(\underline{x})] = \frac{1}{v} - 0 = \frac{1}{v} \tag{175}$$

We shall discuss the variance of $g_1^*(\underline{x})$ later on.

(ii) The gamma distribution. Figure 3 above suggests that we may approximate $h_0(x)$ by a continuous asymmetric distribution. In Appendix III.10 it is shown than an appropriate distribution is the gamma distribution, described in Fisz (1967, pp. 151-154), with parameters $p = 2$ and $b = 2/v$. The resulting estimator is denoted by $g_2^*(\underline{x})$.

(iii) The exponential distribution with modified parameter. Kahn and Marshall (1953, p. 272) remark that for the sake of simplicity $h_0(x)$ may be approximated by the old density function $f(x)$ of (154) but with new values for its parameters.[30] It is simple to derive that the parameter of the exponential distribution should be taken equal to $1/v$ instead of the original value λ. Denote the resulting estimator by $g_3^*(\underline{x})$.

Summarizing, we assumed that we do not know $\xi(\lambda,v)$ and that we therefore estimate ξ. Consequently in the derivation of the three approximations to $h_0(x)$ we should treat $\xi(\lambda,v)$ as unknown. Therefore we cannot choose the parameters of the three approximations such that that the variance is minimized. Instead we so choose these parameters that the range is minimized. Once we have fixed these parameters we cannot select the approximation among the three resulting importance sampling estimators that yields the estimator with the smallest variance. For in Kleijnen (1968b, pp. 186, 187, 198) it is shown that such a selection would imply knowledge of $\xi(\lambda,v)$.[31] This, then is the position in which an experimenter would find himself if he used the Monte Carlo method for the estimation of $\xi(\lambda,v)$. Actually we do know the analytic solution of $\xi(\lambda,v)$; it was presented in (1.2) of Appendix I.1. Using this analytical solution we can calculate the variance of each of the four estimators, $g(\underline{x})$ and $g_i^*(\underline{x})$ $(i = 1, 2, 3)$. In Kleijnen (1968b, pp. 195-198) we derived expressions for the variance of $g(\underline{x})$, $g_1^*(\underline{x})$ and $g_2^*(x)$,[32] while in Appendix III.9 we derive the variance of $g_3^*(\underline{x})$; all these expressions involve $\xi(\lambda,v)$. This yields Table 1 below.[33]

Table 1 shows that $g_1^*(\underline{x})$ based on the shifted exponential distribution leads to large variance reductions. The other two approximations may also yield considerable reductions.[34] These variance reductions need correction for the extra computer time importance sampling requires. For in importance sampling we have to evaluate a more complicated function $g^*(x)$ instead of $g(x)$. Moreover for $g_2^*(x)$ which is based on the gamma distribution we need to generate two exponential variates to obtain one gamma variate as we explain in Appendix III.8. Using rough estimates of the extra computer time, it is shown in Kleijnen (1968b, p. 191, Table 2) that $g_2^*(\underline{x})$ and $g_3^*(\underline{x})$ loose much of their efficiency; however, $g_1^*(\underline{x})$ still yields estimated corrected variances ranging from 3 to 17% of that of $g(\underline{x})$.

TABLE 1. The Variances of the Three Importance Sampling Estimators $g_i^*(\underline{x})$ $(i = 1, 2, 3)$ as a Percentage of $\text{var}[g(\underline{x})]$

		$\text{var}[g_i^*(\underline{x})]/\text{var}[g(\underline{x})] \times 100\%$		
λ	v	$i = 1$	$i = 2$	$i = 3$
0.5	4	0.9%	31.4%	50.4%
0.5	3.5	1.4%	37.5%	59.9%
0.5	2	6.5%	76.6%	100.0%
0.25	8.5	0.7%	28.8%	46.1%

We conclude this section with some more references concerning the use of importance sampling in Monte Carlo studies. Clark (1961, pp. 610-614) gives an elementary discussion of the use of importance sampling for the estimation of

$$\int_{-\infty}^{1} \lambda e^{-\lambda x}\, dx \tag{176}$$

Similar examples can be found in Haitsma and Oosterhoff (1964, pp. 29-30), Hillier and Lieberman (1968, pp. 458-460), Kahn and Marshall (1953, p. 270) and Molenaar (1968, p. 121). Further King (1953, pp. 50-51) briefly discusses importance sampling (without using this term) in "random walks"; compare also the example in Kahn and Marshall (1953, p. 272) concerning the estimation of the probability that a nuclear particle passes through a shield. Evans (1963) and Hastings (1970) applied importance sampling as part of their new VRT's. Pugh (1966) developed a sequential procedure for the estimation of the optimal new density function $h_0(x)$. Clark (1961, p. 618) and Kahn and Marshall (1953, p. 271) give examples demonstrating that an inappropriate choice of the new density function can result in a drastic *variance increase*! Kahn and Marshall (1953, p. 273) prove that importance sampling can also reduce the variance of the estimated *variance* of the "response" $g(x)$ in (154). Moreover Kahn (1955, pp. 11-18) and Kahn and Marshall (1953, pp. 271-272) give the optimal new density function $h(x,y)$ if there are not just one but two *stochastic input variables*, i.e., (154) becomes

$$\xi = \int_{-\infty}^{\infty} \int_{-\infty}^{\infty} g(x,y)\, f(x,y)\, dx\, dy \tag{177}$$

An example of (177) is presented by Clark (1961, pp. 614-616). He points out that if there are more inputs and consequently the process is more complicated then it becomes harder to find appropriate new density functions and this makes it more difficult to realize an efficiency gain by importance sampling. This is also illustrated by the example given by Clark (1961, pp. 616-619) regarding a machine with stochastic running times and a certain scrapping probability. Actually this example takes us to the area of simulation which we shall discuss in the next section.

III.5.3. Importance Sampling in Simulation

Handscomb (1968, p. 7) briefly discusses the difference between Monte Carlo and simulation problems and its consequences for importance sampling. His importance sampling procedure adapted for simulation does not seem suited for general application in simulation. His exposé shows how difficult it is to apply importance sampling to simulation. The machine scrapping problem presented by Clark (1961, pp. 616-619) and the queuing system discussed by Ehrenfeld and Ben-Tuvia (1962, pp. 270-271) and Gürtler (1969, pp. 84-93) also demonstrate this difficulty, for a thorough analysis of the system is required before the new density function can be selected. Therefore Moy (1965, 1966) proposes a standard type of new density function for importance sampling in simulation. In this way we are no longer forced to analyze the system thoroughly in order to select the appropriate type of density. We shall study Moy's proposal in the rest of this section.

In simulation we obtain the response of one simulation run, say $\underline{x}$, from a sequence of stochastic input variables which in turn are generated from a sequence of random numbers, $\underline{r}_t$ $(t = 1, 2, \ldots, \underline{m})$. Hence $\underline{x}$ is some function, say g (this function is specified by the computer program) of the random numbers. Let us first assume, as Moy does, that m is fixed. Then (178) holds.

$$\underline{x} = g(\underline{r}_1, \underline{r}_2, \ldots, \underline{r}_m) = g(\vec{\underline{R}}) \tag{178}$$

where $\vec{\underline{R}}$ denotes the vector of m random numbers used in a simulation run. The expected value of $\underline{x}$ is given by (179).

$$E(\underline{x}) = \int_{-\infty}^{\infty} \cdots \int_{-\infty}^{\infty} g(r_1, r_2, \ldots, r_m)\, f(r_1, r_2, \ldots, r_m) dr_1 dr_2 \cdots dr_m$$

$$= \int_{-\infty}^{\infty} \cdots \int_{-\infty}^{\infty} g(\vec{R})\, f(\vec{R})\, d\vec{R} \qquad (179)$$

So (179) is the "multiple" analog of (154) and (155). Since the random numbers $\underline{r}_t$ are independent we can write their joint density function as the product of their individual density functions, say f_t $(t = 1, 2, \ldots, m)$, as we saw in Section I.5, Eq. (4). Further these individual density functions all are the same, say $f_t(r_t) = s(r_t)$. Hence

$$f(\vec{R}) = f(r_1, r_2, \ldots, r_m) = f_1(r_1)\, f_2(r_2) \cdots f_m(r_m)$$

$$= s(r_1)\, s(r_2) \cdots s(r_m) \qquad (180)$$

In Section I.5, Eq. (2), we saw that $s(r_t)$ satisfies

$$s(r_t) = 1 \quad \text{for } 0 \le r_t \le 1$$
$$= 0 \quad \text{elsewhere} \qquad (181)$$

Combining (180) and (181) yields

$$f(\vec{R}) = f(r_1, r_2, \ldots, r_m) = 1 \quad \text{if all } r_t \text{ are in } [0,1]$$
$$= 0 \quad \text{if any } r_t \text{ is outside } [0,1]$$
$$(182)$$

After substitution of (182) into (179) we obtain

$$E(\underline{x}) = \int_0^1 \cdots \int_0^1 g(\vec{R})\, d\vec{R} \qquad (183)$$

Analogous to (157) we can express (179) as

$$E(\underline{x}) = \int_{-\infty}^{\infty} \cdots \int_{-\infty}^{\infty} \frac{g(\vec{R})\ f(\vec{R})}{h(\vec{R})}\ h(\vec{R})\ d\vec{R} \tag{184}$$

Or using the less general expression (183) we can write

$$E(\underline{x}) = \int_{0}^{1} \cdots \int_{0}^{1} \frac{g(\vec{R})}{h(\vec{R})}\ h(\vec{R})\ d\vec{R} \tag{185}$$

Hence, if $h(\vec{R})$ is a (joint) density function then we can sample the r_t from $h(\vec{R})$ and use the importance sampling estimator, say $\underline{z}$ in (186) which is analogous to (158).

$$\underline{z} = g^*(\vec{R}) = g(\vec{\underline{R}})\ f(\vec{\underline{R}})/h(\vec{\underline{R}}) = \underline{x}\ f(\vec{\underline{R}})/h(\vec{\underline{R}}) \tag{186}$$

If we use (182) then (186) reduces to

$$\underline{z} = g^*(\vec{\underline{R}}) = g(\vec{\underline{R}})/h(\vec{\underline{R}}) = \underline{x}/h(\vec{\underline{R}}) \tag{187}$$

So the computer program generates the response $\underline{x}$ not from random numbers, i.e., stochastic variables $\vec{\underline{R}}$ having the joint density $f(\vec{R})$ specified in (182), but from stochastic variables $\vec{\underline{R}}$ having the density $h(\vec{R})$. The resulting response is corrected by the weighing factor $1/h(\vec{\underline{R}})$. We shall use the term "importance numbers" to distinguish stochastic variables with density $h(\vec{R})$ from (pseudo) random numbers.[35] The variance of the importance sampling estimator is given by

$$\text{var}(\underline{z}) = E(\underline{z}^2) - [E(\underline{z})]^2$$

$$= \int_{-\infty}^{\infty} \cdots \int_{-\infty}^{\infty} \frac{g^2(\vec{R})\ f^2(\vec{R})}{h^2(\vec{R})}\ h(\vec{R})\ d\vec{R} - [E(\underline{x})]^2 \tag{188}$$

that may be further simplified. It is easily checked that taking $h(\vec{R})$ in (188) equal to

$$h_0(\vec{R}) = \frac{g(\vec{R})\ f(\vec{R})}{E(\underline{x})} \tag{189}$$

reduces $\text{var}(\underline{z})$ to zero. So (189) gives the <u>optimal</u> new density function, provided that $g(\vec{R}) \geq 0$ for all $\vec{R}$ since otherwise (189) would result in negative densities. The condition $x = g(\vec{R})$ be not negative, seems not very restrictive in simulation studies. Note that (189) is the multivariate analog of (162).

Before continuing our discussion of the new density function $h(\vec{R})$, we shall generalize the definition of the response $\underline{x}$ so that it encompasses the case of a <u>stochastic</u> number of random numbers per run.[36] So (178) is generalized by

$$\underline{x} = g(\underline{r}_1, \underline{r}_2, \ldots, \underline{r}_{\underline{m}}) \tag{190}$$

Hence

$$E(\underline{x}) = \sum_m E(\underline{x} \mid \underline{m} = m)\ P(\underline{m} = m) \tag{191}$$

where Σ runs over all possible values of $\underline{m}$. Denote the joint density function of m random numbers by $f_m(\vec{R})$ and the function g in (190) for $\underline{m} = m$ by g_m. Then (192) results.

$$E(\underline{x} \mid \underline{m} = m) = \int_{-\infty}^{\infty} \cdots \int_{-\infty}^{\infty} g_m(\vec{R})\ f_m(\vec{R})\ d\vec{R} \tag{192}$$

Analogous to (186) we can define $\underline{z}_m$, the importance sampling estimator for a given m, as

$$\underline{z}_m = g_m(\underline{\vec{R}})\ f_m(\underline{\vec{R}})/h_m(\underline{\vec{R}}) \tag{193}$$

where $h_m(\vec{R})$ is a joint density function from which we sample $\underline{\vec{R}}$. From (192) and (193) it follows that

$$E(\underline{z}_m) = E(\underline{x} | \underline{m} = m) \tag{194}$$

Before the simulation run starts we do not know the value m that will be realized for $\underline{m}$. So the expected value of the response of the run must be taken over $\underline{m}$, i.e., the expected value of the response is given by (195) which uses (194) and (191).

$$\underset{m}{E} [E(\underline{z}_m)] = \underset{m}{E} [E(\underline{x} | \underline{m} = m] = E(\underline{x}) \tag{195}$$

Hence a simulation run with importance sampling still gives an unbiased estimator. We observe that $\underline{z}_m$ can indeed be generated even while we do not know m until the end of the simulation run. For, as we shall see later on, the importance numbers are sampled independently, i.e., their joint density function can be written as the product of their individual density functions, say $v_t(r_t)$. Or

$$h_m(\vec{R}) = \prod_{t=1}^{m} v_t(r_t) \tag{196}$$

where each $v_t(r_t)$ is specified as a function that does not contain m. So $\underline{r}_t$ is sampled from $v_t(r_t)$ and is sampled independently of $\underline{r}_{t'}$ $(t' \neq t)$. At the end of the run we can calculate $h_m(\vec{R})$ from (196) and substitute the resulting value into (193) in order to weigh the response $g_m(\vec{R})$. The <u>variance</u> of $\underline{z}_m$ follows from (193) and is given by (197), the analog of (188).

$$\operatorname{var}(\underline{z}_m) = \int_{-\infty}^{\infty} \cdots \int_{-\infty}^{\infty} \frac{g_m^2(\vec{R})\ f_m^2(\vec{R})}{h_m^2(\vec{R})}\ h_m(\vec{R})\ d\vec{R} - [E(\underline{z}_m)]^2 \tag{197}$$

Its expectation over $\underline{m}$ is

$$\underset{m}{E} \operatorname{var}(\underline{z}_m)\ P(\underline{m} = m) \tag{198}$$

Hence analogous to (189) we can take $h_m(\vec{R})$ equal to

$$h_{m0}(\vec{R}) = \frac{g_m(\vec{R})\ f_m(\vec{R})}{E(\underline{x}|\underline{m} = m)} \tag{199}$$

and this reduces $\text{var}(\underline{z}_m)$ to zero and therefore reduces (198) to zero. Summarizing, the approximation for the optimal new density function can be based on the result for a non-stochastic m given in (199), or on its equivalent in (189). Next we shall see how Moy derived such an approximation.

From (189) or (199) we see that in order to specify $h_0(\vec{R})$ we would need to know the expected value of the response and the explicit form of the function g, which actually stands for a complicated computer program. Hence we have to approximate $h_0(\vec{R})$. Moy (1965, p. 107) and (1966, p. 21) proposes constructing the computer program such that the response tends to increase with increasing random numbers. We have demonstrated in (43) and (44) above how this characteristic can be realized. Such a characteristic implies that (200) may be assumed to hold at least approximately.

$$g(\vec{R}_a) < g(\vec{R}_b) \tag{200}$$

where $\vec{R}_a$ and $\vec{R}_b$ stand for a vector of m a's and m b's, respectively, $0 \leq a < b \leq 1$. From (182) it follows that

$$f(\vec{R}_a) = f(\vec{R}_b) = 1 \tag{201}$$

Hence substitution of (200) and (201) into (189), and supposing $E(\underline{x}) > 0$, yields

$$h_0(\vec{R}_a) < h_0(\vec{R}_b) \tag{202}$$

Assuming that we sample the importance numbers independently so (196) holds, we see that (202) is satisfied if the individual or marginal new density functions v_t satisfy

$$v_t(a) < v_t(b) \quad \text{for } a < b \tag{203}$$

Therefore Moy decided to have the new density functions $v_t(r_t)$ be increasing functions of r_t. These increasing functions replace the old uniform density functions $s(r_t)$ of (181).

Obviously there are many increasing density functions. We like to select a function that yields a large variance reduction but that does not require much extra computer time for sampling from it and for calculating the weight $f(\vec{R})/h(\vec{R})$. Moreover, as we shall see in a moment, the function should have a form such that its optimal parameters can be estimated without much extra work. Moy (1965, p. 107) investigated several functions. Based on his experimental results Moy (1965, p. 119) decided to use only the following new density function,[37] which is also given in Moy (1966, p. 22):

$$v_t(r_t) = \alpha_t^{r_t}(\ell n\ \alpha_t)/(\alpha_t - 1) \qquad \text{for} \quad 0 \le r_t \le 1 \quad (\alpha_t > 1)$$
$$= 0 \qquad \text{elsewhere} \tag{204}$$

Once we have specified the form of the approximation to the optimal new density function, we have to quantify the parameters, i.e., the α_t. In the Monte Carlo example of Section III.5.2 we quantified the parameters by minimizing the range (since we could not minimize the variance) but in complicated simulation studies we cannot minimize the range. Moy (1965, pp. 109-113) proposes the following approach.

Reduce the number of parameters in the new density function by keeping α_t constant, i.e., put $\alpha_t = \alpha$ in (204). The value of α should be so selected that the variance of $\underline{z}$, the importance sampling estimator, is minimal. The variance of $\underline{z}$ was given in (188) above, where only $h(\vec{R})$ depends on α. So α should be so chosen that (205) holds.[38]

$$\frac{\partial\ \text{var}(\underline{z})}{\partial\alpha} = \int_{-\infty}^{\infty}\cdots\int_{-\infty}^{\infty} g^2(\vec{R})\ f^2(\vec{R})\ \frac{\partial[1/h(\vec{R},\alpha)]}{\partial\alpha}\ d\vec{R}$$

$$= \int_{-\infty}^{\infty}\cdots\int_{-\infty}^{\infty} g^2(\vec{R})\ f^2(\vec{R})\ \frac{1}{h^2(\vec{R},\alpha)}\left[-\frac{\partial h(\vec{R},\alpha)}{\partial\alpha}\right] d\vec{R}$$

$$= \int_{-\infty}^{\infty} \cdots \int_{-\infty}^{\infty} \frac{g^2(\vec{R})\ f(\vec{R})\ \left[-\ \frac{\partial h(\vec{R},\alpha)}{\partial \alpha}\right]}{h^2(\vec{R},\alpha)}\ f(\vec{R})\ d\vec{R} = 0 \tag{205}$$

The integral in (205) can be interpreted as defining the mean of

$$W(\vec{\underline{R}},\alpha) = \frac{g^2(\vec{\underline{R}})\ f(\vec{\underline{R}})\ \left[-\ \frac{\partial h(\vec{\underline{R}},\alpha)}{\partial \alpha}\right]}{h^2(\vec{\underline{R}},\alpha)} \tag{206}$$

where $\vec{\underline{R}}$ has the density $f(\vec{R})$. So (205) implies that the mean of $W(\vec{\underline{R}},\alpha)$ be zero. Since we cannot solve (205), Moy suggests its <u>sample</u> equivalent, i.e.,

$$\overline{W}(\vec{\underline{R}},\alpha) = \sum_{i=1}^{n_1} W(\vec{\underline{R}}_i,\alpha)/n_1 = 0 \tag{207}$$

where $\vec{\underline{R}}_i$ denotes the vector of random numbers generated from $f(\vec{R})$ in the i'th simulation run $(i = 1, \ldots, n_1)$. We can use $f(\vec{R}) = 1$ and (204) with $\alpha_t = \alpha$, i.e.,

$$h(\vec{R},\alpha) = \prod_{t=1}^{m} \alpha^{r_t}(\ell n\ \alpha)/(\alpha - 1)$$

$$= \alpha^{\Sigma_1^m r_t}\ [(\ell n\ \alpha)/(\alpha - 1)]^m \tag{208}$$

Then Moy(1965, p. 112) derived that (207) reduces to

$$\sum_{i=1}^{n_1} \frac{(\Sigma_j\ \underline{r}_{ij})(\ell n\ \alpha)(\alpha - 1) + \underline{m}_i(\alpha - 1 - \alpha\ \ell n\ \alpha)}{\alpha^{\Sigma_j \underline{r}_{ij}}}\ \underline{x}_i^2 = 0 \tag{209}$$

where Σ_j denotes $\Sigma_{j=1}^{\underline{m}_i}$ and $\underline{x}_i = g(\vec{\underline{R}}_i)$ denotes the response of the i'th run. So α should be calculated by (numerical) solution of (209). In the remaining n_2 $(= n - n_1)$ runs we can then apply importance sampling using the new density function (208) with the estimated optimal parameter $\hat{\underline{\alpha}}$ derived from (209).

An important variant of the above procedure also applied by Moy (1965, pp. 117, 122-123) keeps α_t in (204) constant <u>per type of input variable</u>. So if we have K types of inputs then (205) is replaced by

$$\frac{\partial \operatorname{var}(\underline{z})}{\partial \alpha_k} = 0 \qquad (k = 1, 2, \ldots, K) \tag{210}$$

Instead of (207) we now have to solve K equations. This involves much extra computer time. Therefore Moy (1965, p. 125) suggests using a new density function only for that type of event that is expected to have the greatest impact on the response. Or

$$h(\vec{R},\alpha) = f(\vec{R}_2, \vec{R}_3, \ldots, \vec{R}_K)\, h_1(\vec{R}_1,\alpha) \tag{211}$$

where the total vector of random numbers, $\vec{R}$, is split into K parts corresponding with the K types of input and only the first type of input is sampled from a new density function. The optimal α can again be derived from (205). An interesting subject for further research would be the investigation of the applicability of Pugh's "gradient technique" in complex simulations. Pugh (1966) applied the <u>steepest descent</u> method for the <u>sequential</u> estimation of the optimal new density function.

Moy applied the various types of distorted density functions to several systems. Applying the system-independent variant of (208) to single-server systems, Moy (1965, p. 118) and (1966, p. 29) found variance reductions ranging from 33% to 54%. He estimated the optimal parameter α_0 from 175 observations, i.e., n_1 in (209) is 175. Additional experiments performed by Moy (1965, pp. 126-128) indicate that the variance reduction is not very sensitive to fluctuations in the estimate of α_0 so a smaller number of runs can be used to estimate α_0. Application of the same system-independent variant to the complex <u>truck dock</u> system did not yield a variance reduction; instead a variance increase resulted. Using a separate new density function for each type of input, resulted in variance reductions of 43 and 57% for the two variants of this complex system. Applying a new density function only for one input variable as in (211) above, yielded a variance reduction of 30 and 39% for the two variants of the truck dock system. Again the parameters are estimated from a large number of runs (180 or 360 runs).

The experiments by Moy (1965, p. 129) show that the estimated optimal parameter $\hat{\alpha}_0$ remains quite constant for the various types systems, the value being roughly 1.12. Therefore he suggests starting the simulation not with random numbers, i.e., $f(\vec{R}) = 1$, but with importance numbers having the density $h(\vec{R})$ of (208) with $\alpha = 1.12$. (If there are many types of inputs then only the most important type or types are generated from $h(\vec{R})$.) We would add that if we have simulated a related system before, then we may instead estimate α_0 from the runs of this system using an equation like (209). So in the first stage we obtain a number of runs applying a distorted density function $h(\vec{R},\alpha)$ where α is 1.12 or α is estimated from previous runs of a related system. Next we can estimate α_0 from the runs of the first stage and use this estimate of α_0 in the second stage. (We point out that in this approach $f(\vec{R})$ in (206) is no longer equal to one but denotes the distorted density function used in the first stage.)

Both the procedure, revised as in the above paragraph, and the original procedure are two-stage approaches. For in the first stage runs are generated from $f(\vec{R}) = 1$ or from $h(\vec{R},\alpha)$, α being equal to 1.12 or being estimated from previous runs of a related system. Then α_0 is estimated from the, say, n_1 runs of stage 1. Next n_2 $(= n - n_1)$ additional runs are generated using importance sampling with the latest estimate of α_0. If the first n_1 runs yield an average response $\bar{x}_1$ and the next n_2 runs an average response $\bar{x}_2$, then an unbiased estimator of the mean response is given by

$$\bar{x} = w\bar{x}_1 + (1 - w)\bar{x}_2 \tag{212}$$

where w is the weight of the average of the first group of runs. The variance of the weighted average $\bar{\underline{x}}$ is given by

$$\mathrm{var}(\bar{\underline{x}}) = w^2 \mathrm{var}(\bar{\underline{x}}_1) + (1 - w)^2 \mathrm{var}(\bar{\underline{x}}_2) + 2w(1 - w) \mathrm{cov}(\bar{\underline{x}}_1, \bar{\underline{x}}_2) \tag{213}$$

where the covariance term is explained by the fact that $\bar{\underline{x}}_2$ is the importance sampling estimator depending on the estimator of the optimal parameter α_0 which in turn is estimated from the first n_1 runs yielding $\bar{\underline{x}}_1$. We can estimate $\mathrm{var}(\bar{\underline{x}}_1)$ and $\mathrm{var}(\bar{\underline{x}}_2)$ from the n_1 and n_2 replications respectively, but we cannot estimate $\mathrm{cov}(\bar{\underline{x}}_1, \bar{\underline{x}}_2)$ from these replications. Moy (1965, p. 114) conjectured that this covariance is so small that it can be neglected and he performed an experiment to check this conjecture. He estimated the correlation between $\bar{\underline{x}}_1$ and $\bar{\underline{x}}_2$ from 67 observations, $\bar{\underline{x}}_1$ and $\bar{\underline{x}}_2$ being the average of 25 runs of a simple queuing system. The resulting correlation coefficient of 0.101 was tested and found to be insignificant; see Moy (1965, p. 132). If we neglect the covariance in (213) then it is simple to derive that the optimal weight is

$$w_0 = \frac{\mathrm{var}(\bar{\underline{x}}_2)}{\mathrm{var}(\bar{\underline{x}}_1) + \mathrm{var}(\bar{\underline{x}}_2)} = \frac{n_1}{n_1 + n_2 \mathrm{var}(\underline{x}_{1i})/\mathrm{var}(\underline{x}_{2j})} \tag{214}$$

where $\underline{x}_{1i}$ and $\underline{x}_{2j}$ are run i and j in groups 1 and 2, respectively. Moy (1965, p. 115) and (1966, p. 27) mentions the possibility of using the estimated variances to estimate this optimal weight. We observe, contrary to Moy, that such an estimated weight leaves the weighted estimator of the mean unbiased. For

$$E(\bar{\underline{x}}) = \underset{\underline{w}_0}{E} [E(\bar{\underline{x}} | w_0)] = \underset{\underline{w}_0}{E} [\mu_x] = \mu_x \tag{215}$$

The variance, is given by (2.16), the first equality following from (4.1) in Appendix 4.

$$\begin{aligned}
\text{var}(\bar{\underline{x}}) &= \underset{\underline{w}_0}{E}\,[\text{var}(\bar{\underline{x}}|w_0)] + \underset{\underline{w}_0}{\text{var}}[E(\bar{\underline{x}}|w_0)] \\
&= \underset{\underline{w}_0}{E}\,[w_0^2\,\text{var}(\bar{\underline{x}}_1) + (1 - \underline{w}_0)^2\,\text{var}(\bar{\underline{x}}_2) \\
&\quad + 2\underline{w}_0(1 - w_0)\,\text{cov}(\bar{\underline{x}}_1,\bar{\underline{x}}_2)] + \text{var}[\mu_x] \\
&= E(\underline{w}_0^2)\,\text{var}(\bar{\underline{x}}_1) + E[(1 - \underline{w}_0)^2]\,\text{var}(\bar{\underline{x}}_2) \\
&\quad + 2E[\underline{w}_0(1 - \underline{w}_0)]\,\text{cov}(\bar{\underline{x}}_1,\bar{\underline{x}}_2)
\end{aligned} \tag{216}$$

So (216) is the analog of (213), the latter equation being valid for a nonstochastic weight w_0. Even if we neglect the covariance term in (216) the estimator in (217) is biased.

$$\widehat{\underline{\text{var}}}(\bar{\underline{x}}) = \underline{w}_0^2\,\widehat{\underline{\text{var}}}(\bar{\underline{x}}_1) + (1 - \underline{w}_0)^2\,\widehat{\underline{\text{var}}}(\bar{\underline{x}}_2) \tag{217}$$

For $\underline{w}_0$ depends on $\widehat{\underline{\text{var}}}(\bar{\underline{x}}_1)$ and $\widehat{\underline{\text{var}}}(\bar{\underline{x}}_2)$, since $\underline{w}_0$ is defined by the sample analogue of (214). Consequently,

$$E[\underline{w}_0^2\,\widehat{\underline{\text{var}}}(\bar{\underline{x}}_1)] \neq E(\underline{w}_0^2)\,E[\widehat{\underline{\text{var}}}(\bar{\underline{x}}_1)] \tag{218}$$

An analogous relation holds for the second term in (217). Hence estimating the weight w_0 is another source of bias in the estimation of $\text{var}(\bar{\underline{x}})$ (besides the bias caused by the neglection of the covariance term). Therefore we may decide to put $\text{var}(\underline{x}_{1i}) = \text{var}(\underline{x}_{2j})$ in (214); or to estimate w_0 from prior information, or to estimate w_0 from part of the n runs and $\text{var}(\bar{\underline{x}}_1)$ and $\text{var}(\bar{\underline{x}}_2)$ in (217) from the remaining part. Anyhow, the determination of a confidence interval for the mean is difficult. (Remember also that the variances of $\underline{x}_1$ and $\underline{x}_2$ are not the same.)

Moy restricted his investigation of importance sampling to terminating systems, each simulation run yielding a single response. We shall investigate briefly the consequences of dropping these restrictions, following the discussion in Kleijnen (1969, pp. 291-293): If we simulate a nonterminating system then as we saw in Chapter II we may tend to use a few prolonged runs instead of many replicated runs. In importance sampling the first runs are used to estimate the optimal parameter α_0. So, if there are only a few runs, then α_0 may be badly estimated. This decreases the variance reduction (though obviously the response remains unbiased). Regarding the second restriction, if we are interested in more than one output variable, then each run yields several responses. Hence multiple responses mean that we have an equation like (207) from which we estimate the paramctcr α_0 for each output variable. So each output variable would give a different value for $\hat{\alpha}_0$. Nevertheless we want to use a single value for α in $h(\vec{R},\alpha)$. It is questionable whether a compromise α would still yield a variance reduction. If the various output variables react in different directions to the random numbers $\vec{R}$ then it is even difficult to find an appropriate parametric form of $h(\vec{R})$ in the approximation to the optimal density $h_0(\vec{R})$. For we can so write the computer program that waiting times of customers increase with increasing random numbers so (200) above holds. However, this implies that in this example idle time of machines will decrease with increasing random numbers and consequently for this output variable (200) does not hold. Hence it is hard to select a parametric form for $h(\vec{R})$ since we do not even know if we should choose an increasing or decreasing function. Consequently whereas multiple responses do not cause problems for the other variance reduction techniques we have discussed, they lead to conflicting requirements for the new importance density function $h(\vec{R},\alpha)$.

We have seen that importance sampling is a rther sophisticated technique. It requires extra computer time for the estimation of the optimal parameter α from equations like (209). Moreover, extra time is needed for generating importance numbers from purely random numbers (these importance numbers are used to generate the input variables). Finally the weight $f(\vec{R})/h(\vec{R})$ in (186) must be calculated. The extra computer time for importance sampling in the truck dock system can be found in Moy (1966, p. 33). If importance sampling is applied to all four types of input it makes the variance reduction of about 65% decrease to a reduction corrected for extra computer time, of about 43%. If importance sampling is applied to only one type of event in the truck dock system then the reduction decreases from 69 to 67%. In the next sections we shall study two variance reduction techniques that require hardly any extra time.

III.6. ANTITHETIC VARIATES

As we shall see below, the antithetic variates technique tries to create negative correlation between observations, generating one observation from the random number r and the other observation from its "antithetic" partner $(1 - r)$. The technique was invented by Hammersley and Morton (1956) for use in the Monte Carlo estimation of the value of an integral. They estimated the integral applying stratified sampling while sampling from the various strata not independently but with negative correlation. (Compare footnote 13 in Section III.2.4.) For only two strata this negative correlation is created by using the random numbers r and $(1 - r)$. For the use of antithetic variates in Monte Carlo problems we refer to the original publication by Hammersley and Morton and to Hammersley and Mauldon (1956); additional historical references can be found in Andréasson (1972b, p. 2). Its use in Monte Carlo studies is also briefly discussed by Haitsma and Oosterhoff (1964, pp. 34-35), Handscomb (1968, pp. 9-10), Newman and Odell (1971, pp. 64-65) and Tocher (1963, pp. 117-118). A very simple variant of antithetic variates is applied to a Monte Carlo problem by Hillier and Lieberman (1968, p. 461). The application of antithetics when estimating a complete distribution rather than only the mean, is discussed by Burt et al. (1970, p. 442).

In 1958 Harling (1958, pp. 12-13) suggested using this variance reducing technique in simulation studies too. However, he did not elaborate on how to realize this. Ehrenfeld and Ben-Tuvia (1962, pp. 273-274) suggested how the antithetic technique can be applied in a simple queuing system but their technique seems not generally applicable in simulation. Finally Tocher (1963, pp. 173-175) proposed a simple expedient for the creation of negative correlation between two "observations," i.e., between two replicated runs in the simulation of a complex system. He suggests generating one run from the random numbers $r_1, r_2, \ldots$ and the other, compagnon or antithetic run from the random numbers $(1 - r_1), (1 - r_2), \ldots$. Remember that in Section I.6.4 Eq. (19), we proved that $(1 - \underline{r})$ is indeed a random number too. Tocher assumed that this procedure does create negative correlation between the two responses. We shall return to this assumption in a moment. Negative correlation between the two responses is desirable since it decreases the variance of the estimated response. For μ, the mean response of the system, is estimated by

$$\bar{\underline{x}} = \frac{1}{2}(\underline{x}_1 + \underline{x}_2) \tag{219}$$

where $\underline{x}_1$ and $\underline{x}_2$ are the response of runs 1 and 2, respectively. So the variance of the estimator $\bar{\underline{x}}$ is given by

$$\mathrm{var}(\bar{\underline{x}}) = \frac{1}{4}\{\mathrm{var}(\underline{x}_1) + \mathrm{var}(\underline{x}_2) + 2\ \mathrm{cov}(\underline{x}_1, \underline{x}_2)\} \tag{220}$$

Hence the variance of $\bar{\underline{x}}$ decreases if $\underline{x}_1$ and $\underline{x}_2$ are negatively correlated, or equivalently if $\mathrm{cov}(\underline{x}_1, \underline{x}_2)$ is negative. Next we shall investigate why it is assumed that the use of the sequence of random numbers $(1 - r_1), (1 - r_2), \ldots$ creates negative correlation.

(i) For the time being let us suppose that the response $\underline{x}$ depends on a single value of one stochastic input variable $\underline{y}$ (instead of a sequence of various stochastic inputs as it is actually the case in simulation). Further, we shall suppose that $\underline{x}$ is a monotonic function of $\underline{y}$, say a monotonic increasing function g_1 of $\underline{y}$, i.e.,

$$g_1(y_1) > g_1(y_2) \quad \text{if} \quad y_1 > y_2 \tag{221}$$

We know that if $\underline{y}$ is generated from the random number $\underline{r}$ by the inversion technique defined in Eq. (13) of Section I.6.4, then $\underline{y}$ is a monotonic increasing function, say g_2, of $\underline{r}$.[39] If $\underline{y}$ is not generated by the inversion technique then such a type of function usually holds too.[40] Hence

$$g_2(r_1) > g_2(r_2) \quad \text{if} \quad r_1 > r_2 \tag{222}$$

Consequently the response $\underline{x}$ is a monotonic increasing function, say g_3, of the random number $\underline{r}$, i.e.,

$$g_3(r_1) > g_3(r_2) \quad \text{if} \quad r_1 > r_2 \tag{223}$$

Hence high values x are generated from high values r. But a high value r gives a low value for its antithetic partner $(1 - r)$ and this in turn yields a low value x. So "high" values (i.e., values higher than the expected value μ) of $\underline{x} = g_3(\underline{r})$ go together with "low" values (i.e., values smaller than the expected value μ) of the antithetic response, say $\underline{x}_a = g_3(1 - \underline{r})$ and this means that $\underline{x}$ and $\underline{x}_a$ are negatively correlated. (The index a refers to antithetics.) For a more formal proof of this negative correlation we refer to Andréasson (1972b, p. 4). Both responses have the same <u>mean</u> since both are generated from a stochastic variable with the same distribution. Or more formally, we have

$$E(\underline{x}) = \int_{-\infty}^{\infty} g_3(r)\ f(r)\ dr = \int_0^1 g_3(r)\ dr \tag{224}$$

and denoting the antithetic random number $(1 - r)$ by $\underline{z}$ we have

$$E(\underline{x}_a) = \int_{-\infty}^{\infty} g_3(z)\ f(z)\ dz = \int_0^1 g_3(z)\ dz \tag{225}$$

where $\underline{z}$ has the same density function f as $\underline{r}$ has, as we proved in Section I.6.4, Eq. (19).

Examples of <u>nonmonotonic</u> relations between the response $\underline{x}$ and the random number $\underline{r}$ are given in the Fig. 4a, 4b. If we knew the value c then we could define new antithetic random numbers, say $\underline{r}_2$, where

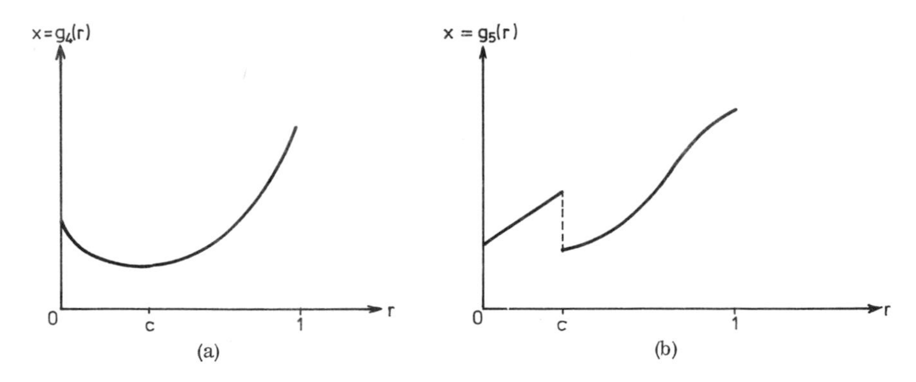

Fig. 4. Nonmonotonic relations between the response $\underline{x}$ and the random number $\underline{r}$.

$$\underline{r}_2 = c - \underline{r} \quad \text{if} \quad 0 \leq \underline{r} \leq c$$
$$= 1 + c - \underline{r} \quad \text{if} \quad c < \underline{r} \leq 1 \tag{226}$$

It is simple to prove that $\underline{r}_2$ is indeed a random number and yields a response that is negatively correlated with $\underline{x} = g_4(\underline{r})$ and $\underline{x} = g_5(\underline{r})$; see also Kleijnen (1968a, p. 28). Unfortunately, in the simulation of a complex system we do not know the value c so we cannot apply (226). Therefore we assume that there is indeed a monotonic relation between the response and the random number.[41)] This assumption seems justified in many applications, e.g., in queuing systems where the waiting time of customers increases with decreasing interarrival times and increasing service times. In these queuing systems, however, we do not have a single stochastic input but a sequence of inputs; moreover, we have inputs of various types. Next we shall study these complications.

(ii) In simulation the response depends on a sequence of values of the stochastic input variable, i.e., instead of $\underline{x} = g_3(\underline{r})$ we have[42)]

$$\underline{x} = g_6(\underline{r}_1, \underline{r}_2, \ldots, \underline{r}_m) \tag{227}$$

We can also reason as follows. Let $\underline{x}$ be the average response (e.g., average waiting time per customer) over the simulated period, i.e.,

$$\underline{x} = \sum_{j=1}^{m} \underline{x}_j/m \tag{228}$$

So the covariance between $\underline{x}$ and the response $\underline{x}_a$ of its antithetic companion run, is given by

$$\text{cov}(\underline{x},\underline{x}_a) = \sum_{j=1}^{m} \sum_{h=1}^{m} \text{cov}(\underline{x}_j,\underline{x}_{ah})/m^2 \tag{229}$$

where $\underline{x}_{ah}$ denotes the h'th individual response in the antithetic run. None of the m^2 covariances in (229) can be assumed a priori to be zero, since for each j and h we know that $\underline{x}_j$ and $\underline{x}_{ah}$ partly use the same sequence of random numbers. What we could assume is that the individual observations $\underline{x}_j$ and $\underline{x}_{ah}$ strongly depend on the most current value of the random numbers, i.e.,

$$\underline{x}_j = g_7(\underline{r}_j) \tag{230}$$

However, this would mean that we "assume away" the dependence between successive individual observations so we can use the reasoning followed in (i) above. We conclude that in the simulation of complex systems it seems impossible to show analytically that antithetic variates lead to a negative correlation between runs. So it is not surprising that Tocher (1963, p. 174) leaves open the possibility that "in some way the assumption of negative correlation is unjustified" and Page (1965, p. 303) observes "how far these considerations lead to large negative correlations between the new observations [$\underline{x}_{ah}$] and those of the basic simulation [$\underline{x}_j$] can only be found by trial." Andréasson (1972b, pp. 4-6, 15-30) showed analytically that in the single-server queuing system antithetics indeed yield negative correlation. We also refer to Radema (1969, pp. 18-20). We shall present the results of the application of antithetic variates to several systems in order to demonstrate that the desirable negative correlation is indeed created. First, however, we shall discuss one more complication, namely the existence of more than one type of input variable.

(iii) Suppose we have K types of stochastic input variables, e.g., service times, interarrival times, etc. Tocher (1963, pp. 174, 175) suggests that if an antithetic run is to be made then one input variable should be generated antithetically and "all other generations from other distributions are started at the same initial value in the two runs...It is best not to change more than one variable at a time since a second change may weaken the correlation caused by the first one." He further suggests a "$\underline{2^K}$ <u>factorial experiment</u>" approach, i.e., if K is, say, 3 then the number of runs is a multiple of $2^K = 2^3 = 8$, each set of eight runs being generated in accordance with Table 2. In Table 2 a plus sign in the row of variable k means that this variable is generated using the same random numbers as were used for this variable in the other runs of the set; a minus sign means that this variable is generated antithetically. For example, comparing the runs 1 and 2 in the set of eight runs shows that input variable 1 is generated antithetically while input variable 2 is generated from the same random numbers in both runs, and input variable 3 is also sampled using the same random numbers in both runs. The runs in a particular set are independent of those in another set since new random numbers are taken for each set.

TABLE 2. The 2^K Factorial Experiment Approach for Generating Antithetic Runs if K = 3[a]

	Number of the run in a set							
Input variable	1	2	3	4	5	6	7	8
1	+	-	+	-	+	-	+	-
2	+	+	-	-	+	+	-	-
3	+	+	+	+	-	-	-	-

[a](-) denotes antithetics, (+) denotes common random numbers.

To see the consequences of Tocher's approach we shall first consider a system with many types of stochastic input variables, i.e., K is high. We look only at the runs 1 and 2 in the set of 2^K runs. Then Tocher's approach implies that a single input variable is generated antithetically, while all remaining (K - 1) input variables are generated using the same vectors of random numbers $\vec{R}_{k'}$ (k' = 2, 3, ..., K) as in run 1. Hence, unless input variable 1 has a completely dominating influence on the output, we expect that the response of run 2 is close to that of run 1. Therefore we expect that the runs 1 and 2 will be positively instead of negatively correlated. Next as an example of a system with a low value of K we consider the single-server queuing system. If in run 1 we happen to generate many small values of input 1 (interarrival time) and many high values of input 2 (service time) then we obtain a high average waiting time. If, contrary to Tocher's suggestion, we take both inputs antithetically in run 2 then we obtain many large values for the interarrival times and many small values for the service times. Hence a low average waiting time results. So this example again demonstrates that it is better not to follow Tocher's approach (but instead to generate all input variables antithetically in the compagnon run). Further, the simplest example of a "system" with two types of input variables is obviously a response depending on only one value of each input variable, i.e., $\underline{x} = g(\underline{r}_1, \underline{r}_2)$. For some specific functions g we calculated the correlation that results if we take either both inputs antithetically or only one input antithetically. These simple examples showed that taking both inputs antithetically gives the strongest negative correlation. For further comments on Tocher's "2^K factorial" approach we refer to Andréasson (1972b, pp. 9-10), Kleijnen (1968a, pp. 4-7) and Radema (1969, pp. 16-18). Radema found analytically and experimentally that in a single-server queuing system Tocher's approach gives bad results. Andréasson (1971b, p. 8) used identical arrival and antithetic service times in his simulation of a simple telephone queuing system; the resulting correlation between two partner runs was found to be positive instead of negative. (In private communication Tocher agreed with the conclusion that the 2^K factorial approach does not work, though some disagreement remained about the exact cause of this failure.)

Summarizing (iii) we suggest that instead of 2^K correlated runs we generate pairs of runs. One run is generated from the K vectors of random numbers $\vec{R}_k$ $(k = 1, 2, \ldots, K)$ where $\vec{R}_k$ is used to generate the input variable k and has m_k elements; the other run is generated from the K vectors of antithetic random numbers $(\vec{I}_k - \vec{R}_k)$ where $\vec{I}_k$ denotes a vector of m_k one's. In order to make each type of input variable have its own vector of random numbers, each of the K input variables has its <u>own</u> random number <u>generator</u>. (We shall return to these generators later on in this section.)

For the sake of simplicity we have presented only the case of nonstochastic m_k. In certain systems there may be a stochastic $\underline{m}_k$, so in the antithetic run more or fewer random numbers are needed than in the original run. With Tocher (1963, p. 174) we assume that this has no serious effect though it tends to weaken the correlation. Stochastic $\underline{m}_k$ lead us to the more important problem of "<u>synchronization</u>." If the i'th random number r_i generates a particular event (e.g., arrival of customer j) then in the antithetic run $(1 - r_i)$ should generate the same event (i.e., not the arrival of customer j' where $j' \neq j$ and not a service time). This synchronization increases the desired negative correlation. Synchronization is simplified if each input type k has its own generator. Compare Andréasson (1971b, pp. 7-8) and Emshoff and Sisson (1971, p. 197) for a discussion of synchronization; in the next section on common random numbers we shall return to the synchronization problem and discuss an example.

Next we consider in more detail several <u>techniques</u> for realizing <u>antithetic</u> random numbers and stochastic variables.

(a) Generate the random number r and <u>subtract</u> it from the number 1. This subtraction is the only extra instruction that must be executed for each random number. Compared with the total computer time needed for a run this extra time will be negligible. If we have enough computer memory space available then we may store the original sequence of random numbers. In the antithetic run we have only to calculate $(1 - r)$ without having to generate r again; such a procedure was applied by Andréasson (1971b, p. 7); see also Garman (1971, pp. 9-10).

(b) If the random number generator is <u>multiplicative</u>, i.e.,

$$z_i = az_{i-1} \pmod{m} \qquad (i = 1, 2, \ldots) \tag{231}$$

$$r_i = z_i/m \tag{232}$$

then the antithetic random numbers $(1 - r_i)$ can be obtained simply by making the <u>starting</u> value z_0 "antithetic," i.e., if we take as starting value

$$z_0^* = m - z_0 \tag{233}$$

in (231) then the resulting random numbers, say r^*, are antithetic, i.e., $r_i^* = 1 - r_i$. This is proved in Appendix 10.[43] In this way no additional computer time is needed; see also Andréasson (1972b, pp. 12-13).

(c) If we have a <u>discrete</u> variable then we can apply the procedure illustrated in Fig. 5a, b.[44] Here we switched the order of the values of the variable on the abscissa. Hence a high random number r generates the high value y_3 in Fig. 5a, but the low value y_1 in Fig. 5b. So for the generation of antithetic values of the input variable we need an <u>extra table</u> in computer memory, where the input variable is tabulated in <u>reversed</u> order. The extra time needed to direct the program to one of the two tables is negligible. This provides another procedure for generating antithetic values of discrete input variables.[45]

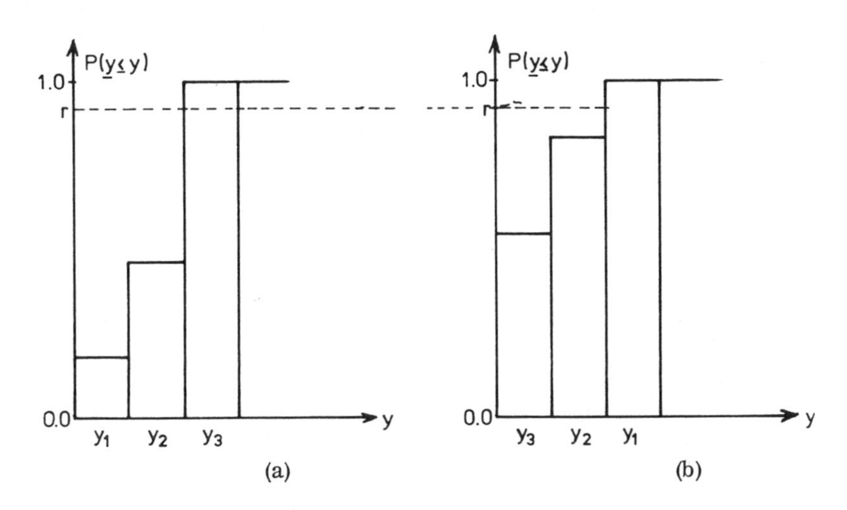

Fig. 5. Antithetic generation of a stochastic variable $\underline{y}$ from the random number $\underline{r}$.

(d) Tocher (1963, p. 16, Eq. (1)) mentions the possibility of approximating the inverse cumulative distribution function F^{-1} in $\underline{y} = F^{-1}(\underline{r})$ by the so-called Russel function, i.e, $\underline{y} = F^{-1}(\underline{r})$ is approximated by

$$\underline{y} = A + B\underline{r} + C\underline{r}^2 + a\underline{r}^2 \log(1 - \underline{r}) + b(1 - \underline{r})^2 \log \underline{r} \quad (234)$$

Farmer (1966, p. 11) states that the antithetic values of $\underline{y}$ are obtained if the coefficients in (234) are replaced by the coefficients

$$A^* = A + B + C, \quad B^* = -(B + 2C), \quad C^* = C, \quad a^* = b, \quad b^* = a \quad (235)$$

However, since the Russel approximation is not much used in simulation, Farmer's procedure does not seem very useful.

(e) Andréasson (1970, pp. 6-7) and (1971a, p. 20) proposes generating antithetic values y_a for the input variable $\underline{y}$ having a symmetric continuous distribution and mean η, by means of the relation

$$y_a = 2\eta - y \tag{236}$$

It is simple to prove that his technique generates the same values as the inversion technique using antithetic random numbers shown in (237).

$$y_a = F^{-1}(1 - r) \tag{237}$$

If the old values y are stored then in general Andréasson's procedure is faster since it requires only a simple subtraction instead of the function F^{-1}. His technique can also be applied when the old values y are not generated by inversion i.e., y is not $F^{-1}(r)$ (e.g., $y = \Sigma_1^{12}\ r_i$ when normal variates are required).

We observed above that in our opinion best results are obtainable when each input variable has its <u>own</u> stream of random numbers. If we have e.g., the multiplicative random number generator defined in (231) and (232) above, then each type of input can be generated with a separate value for the <u>multiplier</u> a in (231). However, in Section I.5 we saw that the multiplier should be chosen very carefully. Therefore we may decide to use only one particular value for the multiplier. Then another possibility is to use a separate <u>starting</u> value z_0 in (231) for each type of input variable. In that approach we have to avoid overlap between the various streams of random numbers. This overlap can be prevented if we have used the random number generator before and have kept a record of the starting and end values of the random number sequences. Mize (1973, pp. 11-13, 17) presents another approach to the creation of multiple random number streams. His approach, however, has doubtful merits in our opinion. Mihram (1972, pp. 246-249) also discusses multiple streams but in a different context, viz., not for variance reduction but for verification.

Another possibility arises in the single-server queuing system where it is known that after each interarrival time a service time is sampled, i.e., $\underline{r}_{2i-1}$ is used to generate an interarrival time and $\underline{r}_{2i}$ is used for a service time $(i = 1, 2, \ldots)$. In this case we do not need two random number generators but we can use one generator for the two types of input. In such a queuing system negative correlation between two runs can also be created by the following procedure. In run 1 the arrival times are generated from the random numbers $\vec{R}_1$ and the service times from $\vec{R}_2$; in run 2 $\vec{R}_1$ and $\vec{R}_2$ are switched, i.e., the arrival times are now generated from $\vec{R}_2$, the service times from $\vec{R}_1$. This technique was first proposed by Marshall.[46)] It was applied by Page (1965) and later by Radema (1969). This procedure cannot be used in complex systems and therefore we shall not discuss it further.

We shall now present some applications of antithetic variates in simulation. Most applications concern the simulation of the well-known single-server queuing system. Antithetics were applied to this system by Lombaers (1968, pp. 252-253), Moy (1965, pp. 72-74, and 1966, pp. 11-14, 27-35), Page (1965), and Radema (1969); see also Gaver (1969). The first three authors found the following average variance reductions where we place the utilization factor of the system between parentheses: 71% (80%), 41% (60 to 100%) and 46% (75 to 90%) respectively. Radema made an investigation of the effect of the utilization factor of the system on the variance reduction. This factor was found to influence the variance reduction strongly. For, if the utilization degree is low then waiting times will be zero even for small interarrival times and high service times, i.e., the monotonic increasing relation in (223) above will be replaced by a relation more like (238).

$$g_3(r_1) = g_3(r_2) = 0 \qquad \text{if} \quad r_1 > r_2 \tag{238}$$

The table in Radema (1969, p. 13) shows that the variance reduction ranges from 5% for a utilization of 20 to 42% for a utilization of 100%. Andréasson (1972b, pp. 15-59) made a massive study on the effects of antithetics (including variants mentioned in footnote 41) for varying utilization factors, input distributions and response quantities. He also derived analytical results for extreme utilization factors (viz., zero or 1) in the single-server queuing system. In general he found that increasing utilization strengthens the variance reduction; however, for some response variables the variance reduction decreases for extreme utilization. Andréasson (1971b, p. 5) further studied a simple telephone queuing system where a particular probability p is to be estimated. He derived analytically the variance reduction that could be realized if the (negative) correlation between the outputs of the two partner runs were minimized. This variance reduction becomes smaller as p approaches one or zero. In practice the extreme correlation between the two outputs was not realized; nevertheless the estimated variance reductions decreased as p approached zero or one; see the diagrams in Andréasson (1971b, pp. 21-22). Radema also investigated the effects that run-length and starting conditions may have on the variance reductions. The antithetic technique was further applied by Kleijnen (1969, p. 291) in a queuing system with four service stations in sequence, where the effect of changing the parameters of the service distributions of these four stations was estimated; the estimated variance reduction was 41%. Shedler and Yang (1971, pp. 122-123) simulated the queuing problems within a multiprogramming computer and found that for one response variable sizable variance reductions resulted but for another response variable variance increases could occur. Gürtler (1969, p. 83) found variance reductions of 5 to 10% in a queuing system with parallel service stations. Andréasson (1970, pp. 8, 17) applied antithetics to two different queuing systems, each system having service stations both in parallel and in sequential order. Variance reductions were roughly 40%; application of antithetics required 11% additional computer time. Moy used antithetics in his complex truck dock system and found variance reductions of 18% to 27%.[47)] Burt et al. (1970) applied antithetics to various simple stochastic networks and, roughly speaking, halved the variances; see also Burt and Garman (1971a, pp. 251-254). Mayne (1966) applied antithetics to a quite different problem (the estimation of the gradient of the cost function in a nonlinear discrete-time control system) and he found that the estimated variance was reduced by a factor of 50 to 500. Garman (1971, pp. 19-20) applied antithetics to an artificial system (viz., an algorithm not meant to reproduce a real-world system) and found 22% variance reduction provided the random number streams were kept synchronized. Many of the above applications are also summarized in Andréasson (1972b, pp. 60-64). Note that in none of the above applications Tocher's 2^K factorial approach was followed.

The statistical analysis of the simulation runs is not complicated by antithetic variates. For we can take the average of each antithetic pair, i.e., take

$$\underline{z}_j = (\underline{x}_{2j} + \underline{x}_{2j-1})/2 \quad (j = 1, 2, \ldots, n/2)(n \text{ even}) \tag{239}$$

where $\underline{x}_{2j}$ is the antithetic partner-run of $\underline{x}_{2j-1}$; all $\underline{x}_{2j-1}$ use new sequences of random numbers, i.e., the $\underline{x}_{2j-1}$ are independent. Hence all $\underline{z}_j$ are independent! So it is simple to estimate e.g., the variance of the average response. For

$$\bar{\underline{x}} = \frac{\sum_{h=1}^{n} \underline{x}_h}{n} = \frac{\sum_{j=1}^{n/2} \underline{z}_j}{n/2} = \bar{\underline{z}} \tag{240}$$

Consequently

$$\hat{\text{var}}(\bar{\underline{x}}) = \hat{\text{var}}(\bar{\underline{z}}) = \sum_{j=1}^{n/2} (\underline{z}_j - \bar{\underline{z}})^2/(n/2 - 1) \tag{241}$$

Confidence intervals like (12) can be determined using (240) and (242) and the t-statistic with $(n/2 - 1)$ degrees of freedom. Hence the power of the t-test is decreased or equivalently the length of the confidence interval is increased, because of the loss of degrees of freedom, but this is compensated by the smaller variance of $\underline{z}$ as compared with that of $\underline{x}$.[48]

Obviously the antithetic variates technique is very attractive since it is easy to apply with only little extra programming and running time. Nevertheless worthwhile variance reductions can be obtained and the statistical analysis is not complicated. The technique can be applied in both terminating and non-terminating systems, either with a single response per run or with multiple responses. The dynamic behavior of the system is not disturbed by the antithetics so this behavior can also be studied.

III.7. COMMON RANDOM NUMBERS

All above VRT's can be used to increase the reliability of the estimated response of a particular system. Usually we simulate more than one system since we want to choose among several systems. In that case we are not interested in the absolute values of the system responses but in the differences among system responses. It is intuitively reasonable to compare the systems under "the same circumstances." Because of this argument we are in favor of simulating non-terminating systems with the same starting conditions. Another implication is that, as far as possible, the same values of the inputs are used, e.g., if the systems 1 and 2 both have stochastic interarrival times then both systems are simulated using the same sequence of random numbers. Statistically common random numbers mean that both system responses are correlated. Now let us consider the variance of the difference between $\underline{x}$, the estimated response of system 1, and $\underline{y}$, the estimated response of system 2. The general expression for this variance is given by

$$\mathrm{var}(\underline{x} - \underline{y}) = \mathrm{var}(\underline{x}) + \mathrm{var}(\underline{y}) - 2\,\mathrm{cov}(\underline{x}, \underline{y}) \tag{242}$$

Hence the variance of the estimated difference is decreased if the covariance term in (242) can be made positive. Such a positive covariance is created by the use of the same random numbers if we assume that both systems react to the stochastic input variables in the same direction. This point is illustrated in Fig. 6. Figure 6a shows a situation where $\underline{x}$ and $\underline{y}$ would be negatively correlated when using the same random numbers. In Fig. 6b $\underline{x}$ and $\underline{y}$ will show

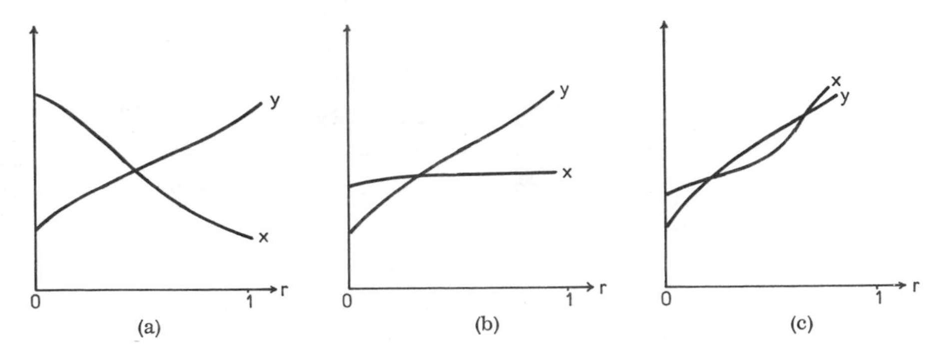

Fig. 6. Simulating two systems with the same random numbers.

only weak correlation, since for most random numbers x will have the same value while y will vary with r. In Fig. 6c where both x and y vary roughly in the same way with the random number r, a positive correlation exists. Of course in simulation there are much more complicated relations between the random numbers and the output. Nevertheless it is reasonable to assume that both systems react in the same direction. For consider the simulation of queuing systems. These systems usually differ from each other in so far as they have different numbers of service stations, mean service rates, mean arrival rates, queuing discipline, etc. Nevertheless these system variants react in the same direction, since low interarrival times and long service times tend to result in longer waiting times whatever the number of service stations, etc. is.

Next we shall consider in more detail what is meant by "using the same random numbers." Obviously this technique implies that each stochastic input variable has its own sequence of random numbers. When simulating the next system variant we start generating input values using the same initial values for the random number generators as in the previous system variant. In a simple queuing system the use of the same initial values for the generators will be sufficient. A simple queuing example showing the effects of separate random number streams per input variable is worked out in Mize (1973). However, as Kosten (1968, p. 8) also remarks, in other systems extra provisions may be necessary to keep the sequence of random numbers in both systems "synchronized." Kahn and Marshall (1953, p. 268) describe this synchronization as "the identical random numbers being used in each case at the same juncture in the realization of the two processes." A more formalized analysis of synchronization is presented in Garman (1971, pp. 15-23) and (1973). He shows that variance reduction decreases as system complexity increases unless synchronization is well taken care of. He proposes not to use multiple random number streams but a single stream where some random numbers are actually not used ("dummies") in order to keep events synchronized; see the end of the following example.

An illustration of the synchronization problem is shown in Fig. 7. In this figure we give the flowchart for the bus maintenance problem that was formulated before in Appendix I.3. (The flowchart in Fig. 7 can be compared with the original flowchart in Fig. 3.1 of Appendix I.3.) The state of the bus at the end of a trip is sampled generating a random number r and using Table 3. Hence a positive correlation between the numbers of trips per day under the two strategies α and β defined in Appendix I.3, may be created as follows. Sample the state of the bus at the end of trip j ($j = 1, 2, \ldots, N$) on day i ($i = 1, 2, \ldots, M$) for both strategy α and β using the same random number. Hence, if r is "small" (say r is smaller than a and b in Table 3) then under strategy α the bus ends the trip in state A and is to be repaired. So one trip is cancelled and this decreases the total number of trips on that day. Under strategy β this random number implies that the

TABLE 3. Sampling the State of the Bus at the End of a Trip

Bus starts trip in state	Bus ends trip in state		
	good	A	B
good	$a \leq r \leq 1$	$0 \leq r < a$	---
A	---	$b \leq r < 1$	$0 \leq r < b$

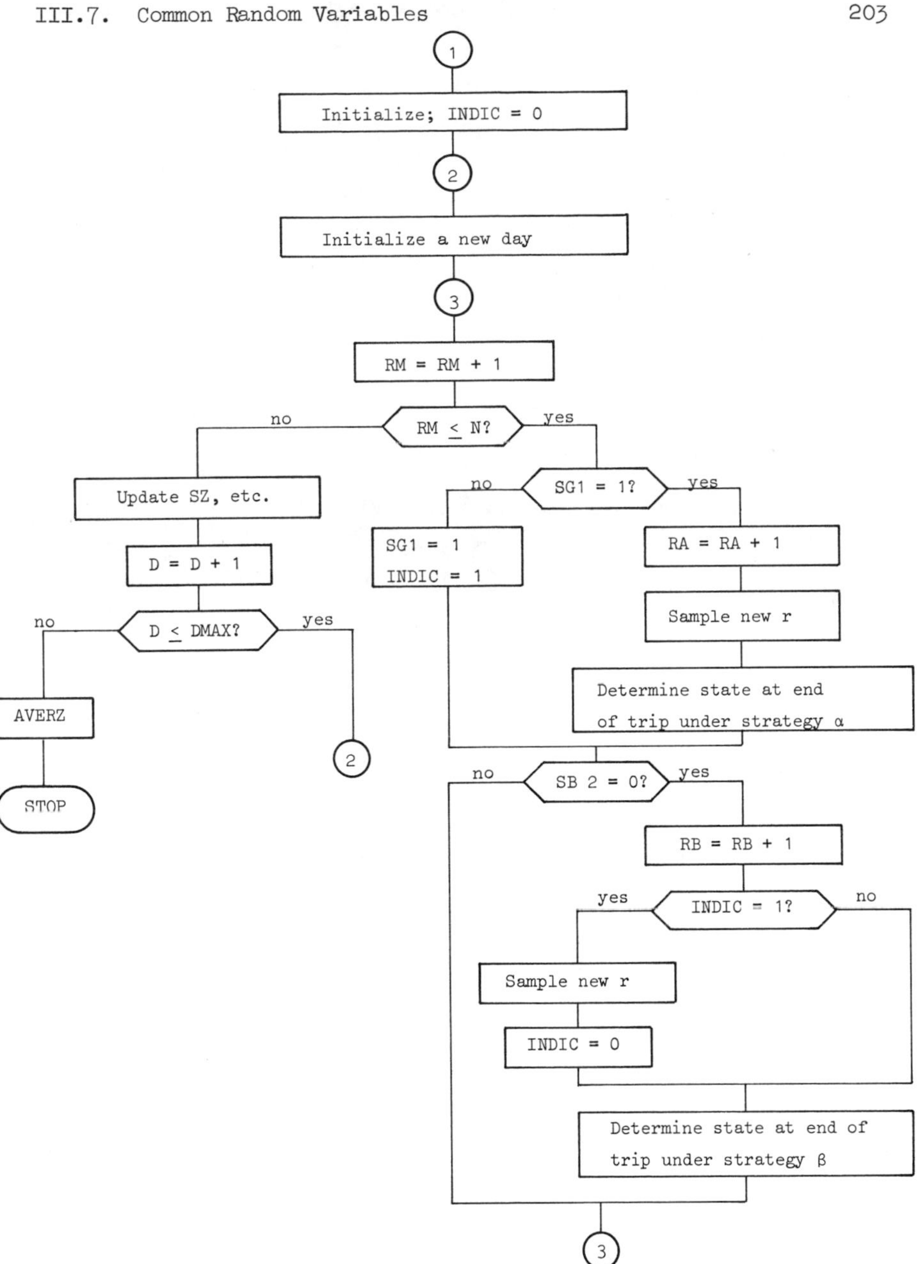

Fig. 7. Simulation of two maintenance strategies using common random numbers.

bus is in state A or B depending on whether it started in state "good" or in state A, respectively. Both events tend to decrease the total number of trips on that day. Summarizing, both strategy α and β tend to yield a low number of trips for low values of the random numbers and this means that positive correlation is created. In order to synchronize the random numbers in strategy α and β we did not first simulate M days of strategy α and next M days of strategy β. Instead we first have the bus make a trip under strategy α and then under strategy β; next the following trip is made, again first under strategy α and then under β, etc. Moreover, we have to avoid the use of the same random number in two consecutive trips under the same strategy, since the model specifies that the state at the end of each trip should be determined using a new random number. Therefore Fig. 7 contains an indicator called "INDIC". This indicator has the value one if under strategy α no new random number is sampled (since the bus is being repaired). Under strategy β a new random number must then be generated which was not used in strategy α. An alternative approach consists of deleting the indicator. Instead a new random number is sampled for strategy α even if the bus is under repair but this random number is then not used. Such a "dummy" random number approach is proposed by Garman (1971, 1973).

Next we shall present some actual variance reductions obtained with this VRT. We simulated the above maintenance system during 1500 days for the parameter values $a = 1/4$, $b = 4/5$, and $N = 25$. The simulation experiment was performed using independent and common random numbers, respectively. Common random numbers yielded an estimated variance reduction of 38%. In Kleijnen (1969, p. 291) we reported that a queuing system with four service stations in sequence gave a variance reduction of 95%. Andréasson (1970, p. 19) realized a 90% variance reduction in his multichannel queuing system. Carter and Ignall (1970, p. 31) found an estimated variance reduction of 73% in their fire department simulation. Brenner (1963, pp. 293-294) estimated how common random numbers or as he calls it "correlated sampling" reduces the "penalty." (Remember that he defined the "penalty" as the true cost at the optimum factor combination estimated by simulation, minus the true cost at the exact optimum determined by the analytical solution.) For his inventory systems Brenner reports considerable penalty decreases when using the same random numbers. Ignall (1972) used common random numbers with response surface methodology in an inventory simulation but he did not quantify the resulting savings. Mize (1973, pp. 13-16) simulated job shops using the same random numbers with or without separate streams per input type; the improved synchronization resulted in smoother estimated cost curves. We could not find any further reports on variance reductions with common random numbers. Nevertheless the use of the same random numbers is briefly mentioned by many authors, e.g., Emshoff and Sisson (1971, p. 197), Farmer (1966, p. 11), Fishman and Kiviat (1967, p. 26), Haitsma and Oosterhoff (1964, p. 34), Maxwell (1965, p. 10), Molenaar (1968, p. 118), Wurl (1971, p. 94), and the other authors mentioned in this section. Mihram (1972, p. 275) implicitly rejects common random numbers (except on p. 401 where he noted that partly using the same numbers creates "blocks" in experimental design terminology).

Obviously the same random numbers can also be used when comparing <u>more</u> than two systems. Further, as we remarked in the section on control variates, a special form of the use of the same random numbers consists of simulating besides the system of interest a simpler system with a <u>known</u> analytical solution, so that (152) above can be applied. For a critical appraisal of this technique we refer back to our discussion in Section III.4.4. Besides using the same random numbers we may even try to run the computer program once only, evaluating several systems at the same time in a single run. Examples are given by Handscomb (1968, pp. 8-9) and Lombaers (1968, p. 254); see also Fig. 7 above. As Radema (1969, p. 16) mentions, if exponential input variates are needed in a simulation then we may store $\ln(r_i)$ instead of r_i $(i = 1, 2, \ldots)$ since the calculation of a logarithm takes much time in a computer. Finally, Conway (1963, p.

53), Conway, Johnson and Maxwell (1959, p. 106) and Hillier and Lieberman (1968, p. 453) observe that the use of the same random numbers implies that we do not have independent observations and this complicates the statistical analysis. Obviously, if we compare only two systems we can calculate the differences

$$\underline{d}_i = \underline{x}_i - \underline{y}_i \qquad (i = 1, 2, \ldots, n) \tag{243}$$

where $\underline{x}_i$ and $\underline{y}_i$ are the response of systems 1 and 2, respectively, and each system is supposed to be simulated the same number of times. Though $\underline{x}_i$ is dependent on $\underline{y}_i$ the new observations $\underline{d}_i$ are mutually independent and therefore it is simple to derive a confidence interval for the average difference $\bar{\underline{d}}$, where

$$\bar{\underline{d}} = \sum_{i=1}^{n} \underline{d}_i/n = \bar{\underline{x}} - \bar{\underline{y}} \tag{244}$$

However, if we are comparing more than two systems, then as we shall explain in Chapter V it is incorrect to use confidence limits derived for the case of two systems; instead we use multiple comparison and multiple ranking procedures. All these procedures assume independent observations. We shall return to the problem of the analysis of correlated observations in Section III.8.7.

Summarizing this section, we see that the use of the same random numbers may yield considerable variance reductions without much extra programming and running time. Its application is very simple and this explains why this technique is the most popular VRT. Actually it is the only VRT that is as a rule used by practitioners of simulation.

III.8. JOINT APPLICATION OF ANTITHETIC VARIATES AND COMMON RANDOM NUMBERS[49)]

III.8.1. Introduction

Moy investigated four VRT's that can be applied to many types of systems. These VRT's are stratified, regression, importance and antithetic sampling. He compared these four techniques by applying them to several variants of the single-server and the truck dock system. Some of his results were given in the preceding sections. For the details of his conclusions we refer to the summary in Moy (1965, pp. 135-139) and (1966, pp. 35-36). If we consider only his results for the complex truck dock system we conclude that the antithetic variates technique is the best system-independent VRT. System-dependent variants of importance sampling and control variates yielded higher variance reductions than antithetic variates. However, these variance reductions should be corrected for the extra computer programming and running time. Hence we would recommend antithetic variates as the simplest effective VRT among the above four techniques. We assume that this technique is so simple that practitioners are willing to apply it. A technique, not investigated by Moy, is the use of the same random numbers. In Section III.7 we saw that this VRT yields considerable variance reduction. Moreover the technique is so simple that it is the only technique with widespread use in practice. Finally, another technique we discussed but that was not investigated by Moy, is selective sampling. In Section III.3.2 we showed that this technique is biased (nevertheless, its MSE may be small) and is not suited for simulation of systems with a stochastic number of random numbers per run. Moreover, it involves extra computer time for the calculation of the desired frequencies v_i^* and the realization of these frequencies by sampling without replacement. (Continuous stochastic input variables need to be grouped into classes before these calculations can be executed.) Summarizing, there are two techniques that are suited for routine application, namely, antithetic variates and common random numbers.

Several authors mention the possibility of applying more than one VRT in a problem. Tocher (1963, p. 103) suggests combining the control variate technique with stratification in his Monte Carlo example. Burt et al. (1970, p. 452) combined antithetic and control variates (resulting in an additional variance reduction of 50%) and so did Andréasson (1971b, p. 12) (but with no additional variance reduction when adding antithetics to control variates). Gaver (1969) found additional variance reduction when combining these two techniques but Gaver and Shedler (1971, p. 448) found little variance reduction. Shedler and Yang (1971) combined antithetics with their stratification technique and found increased variance reduction; see also Fishman (1972) and Lewis (1972, p. 12). Handscomb (1968, p. 4) states that the various techniques "are by no means mutually exclusive."

More specifically, Tocher (1963, p. 177) and implicitly Emshoff and Sisson (1971, p. 198) and Fishman (1967, pp. 23-24) recommend the _combination of antithetic variates and common random numbers_. Nevertheless we shall show that it is not obvious at all that the combination of these two particular methods gives best results.

III.8.2. The Conflict Between Antithetics and Common Random Numbers

At first sight it may look quite reasonable to apply both antithetic variates and common random numbers when estimating the difference between the responses of two systems. For define

$$\underline{\bar{x}} = \sum_{i=1}^{M} \underline{x}_i / M \tag{245}$$

$$\underline{\bar{y}} = \sum_{j=1}^{N} \underline{y}_j / N \tag{246}$$

$\underline{x}_i$ and $\underline{y}_j$ being the response of run i of system 1 and run j of system 2, respectively. Then the difference between the responses of the two systems is estimated by

$$\bar{d} = \bar{x} - \bar{y} \tag{247}$$

so

$$\text{var}(\bar{d}) = \text{var}(\bar{x}) + \text{var}(\bar{y}) - 2\text{cov}(\bar{x},\bar{y}) \tag{248}$$

Antithetic variates reduce the variance of the average of the estimated responses of a particular system, i.e., $\text{var}(\bar{x})$ and $\text{var}(\bar{y})$ are decreased. The use of the same random numbers is supposed to create a positive covariance between $\bar{x}$ and $\bar{y}$. But let us consider the joint application of both techniques in more detail. If we apply both techniques then the runs of systems 1 and 2 are generated in accordance with Table 4. Column (2) of Table 4 shows that system 1 is simulated applying antithetics. ($\vec{1}$ denotes a vector of 1's.) We suppose that this technique indeed creates the desirable negative correlation between x_1 and x_2, between x_3 and x_4, etc. In the same way it follows from column (4) that antithetics are applied to system 2 where they are supposed to create negative correlation between y_1 and y_2, between y_3 and y_4, etc. Further, looking at a particular row in Table 4 we see that the same random numbers are used to run both systems. Hence we suppose that there is indeed positive correlation between x_1 and y_1, between x_2 and y_2, etc. However, Table 4 also shows that there is <u>negative</u> correlation between x_1 and y_2, between x_2 and y_1, between x_3 and y_4, etc. These negative "<u>cross-correlations</u>" are <u>undesirable</u>. (In Section III.8 we use the term "<u>cross-correlation</u>" for correlations between the response x of system 1 and y of system 2; so correlations between runs of the same system are no cross-correlations.) Negative cross-correlations are undesirable since, for instance, a high value of x_1 goes together with a low value of y_2 and this distorts the comparison of the two systems. The effect of the various correlations or covariances is shown more rigorously in (249), this equation being derived in Appendix 11.

TABLE 4. Joint Application of Antithetic Variates and Common Random Numbers

Run	System 1		System 2	
	Random numbers	Response	Random numbers	Response
(1)	(2)	(3)	(4)	(5)
1	$\vec{R}_1$	x_1	$\vec{R}_1$	y_1
2	$\vec{I} - \vec{R}_1$	x_2	$\vec{I} - \vec{R}_1$	y_2
3	$\vec{R}_2$	x_3	$\vec{R}_2$	y_3
4	$\vec{I} - \vec{R}_2$	x_4	$\vec{I} - \vec{R}_2$	y_4
⋮	⋮	⋮	⋮	⋮

$$\begin{aligned} \operatorname{var}(\underline{\bar{d}}) &= M^{-2} \sum_{i \neq g}^{M} \sum^{M} \operatorname{cov}(\underline{x}_i, \underline{x}_g) + M^{-1}\sigma_1^2 \\ &+ N^{-2} \sum_{j \neq h}^{N} \sum^{N} \operatorname{cov}(\underline{y}_j, \underline{y}_h) + N^{-1}\sigma_2^2 \\ &- 2M^{-1} N^{-1} \sum_{i=1}^{M} \sum_{j=1}^{N} \operatorname{cov}(\underline{x}_i, \underline{y}_j) \end{aligned} \tag{249}$$

where

$$\sigma_1^2 = \operatorname{var}(\underline{x}_i) = \operatorname{var}(\underline{x}_g) \qquad i, g = 1, \ldots, M \tag{250}$$

$$\sigma_2^2 = \operatorname{var}(\underline{y}_j) = \operatorname{var}(\underline{y}_h) \qquad j, h = 1, \ldots, N \tag{251}$$

Comparing (249) with (248) we see that the first two terms in (249) correspond with the first term in (248); the next two terms in (249) with the second term in (248) and the last term in (249) with the last term in (248). Further, each run of a system has the same variance so (250) and (251) hold. The magnitude of σ_1^2 and σ_2^2 is determined by the system we are simulating, but the covariances are determined by our choice of the random numbers. From (249) it follows that we should try to choose the random numbers in such a way that the following relations hold.

$$\text{cov}(\underline{x}_i, \underline{x}_g) < 0 \qquad i \neq g; \quad i, g = 1, \ldots, M \tag{252}$$

$$\text{cov}(\underline{y}_j, \underline{y}_h) < 0 \qquad j \neq h; \quad j, h = 1, \ldots, N \tag{253}$$

$$\text{cov}(\underline{x}_i, \underline{y}_j) > 0 \qquad i = 1, \ldots, M; \quad j = 1, \ldots, N \tag{254}$$

When discussing Table 4 we saw that the joint application of both VRT's implies that some cross-correlations, i.e., some $\text{cov}(\underline{x}_i, \underline{y}_j)$, are negative. Hence (254) is violated for some values of i and j. Therefore we shall consider three obvious alternatives in the next section.

III.8.3. Three Alternatives

In this section we shall discuss three alternative methods for the generation of correlated runs, and we shall derive the corresponding variance of the estimated difference between the systems responses.

A. Antithetic Variates Only

For method A Table 4 is replaced by Table 5, where $\vec{P}_j$ denotes a vector of random numbers different from $\vec{R}_i$ ($j = 1, 2, \ldots, N/2$ and $i = 1, 2, \ldots, M/2$). (The symbol P is chosen as a mnemonic for pseudorandom number.)

B. Common Random Numbers Only (in the first N runs)

We leave open the possibility that both systems are not run an equal number of times, i.e., $N \neq M$. If for the time being we call the system with the smallest number of runs "system 2," then $N \leq M$. If $N < M$ then $\text{var}(\bar{\underline{d}})$ obviously decreases if we apply antithetic variates in the last $(M - N)$ runs of system 1. For in that case the last $(M - N)$ runs do not create undesirable negative cross-

TABLE 5. Antithetic Variates Only $(\vec{P} \neq \vec{R})$

	System 1		System 2	
Run	Random numbers	Response	Random numbers	Response
(1)	(2)	(3)	(4)	(5)
1	$\vec{R}_1$	x_1	$\vec{P}_1$	y_1
2	$\vec{I} - \vec{R}_1$	x_2	$\vec{I} - \vec{P}_1$	y_2
3	$\vec{R}_2$	x_3	$\vec{P}_2$	y_3
4	$\vec{I} - \vec{R}_2$	x_4	$\vec{I} - \vec{P}_2$	y_4
⋮	⋮	⋮	⋮	⋮

TABLE 6. Common Random Numbers in the First N Runs; Antithetic Variates in the last (M - N) Runs of System 1. $(\vec{R} \neq \vec{P},\ z = (M - N)/2)$

	System 1		System 2	
Run	Random numbers	Response	Random numbers	Response
(1)	(2)	(3)	(4)	(5)
1	$\vec{R}_1$	x_1	$\vec{R}_1$	y_1
2	$\vec{R}_2$	x_2	$\vec{R}_2$	y_2
⋮	⋮	⋮	⋮	⋮
N	$\vec{R}_N$	x_N	$\vec{R}_N$	y_N
N + 1	$\vec{P}_1$	x_{N+1}		
N + 2	$\vec{I} - \vec{P}_1$	x_{N+2}		
⋮	⋮	⋮		
M - 1	$\vec{P}_z$	x_{M-1}		
M	$\vec{I} - \vec{P}_z$	x_M		

correlations. Therefore we may decide to use common random numbers in the first N runs while the following (M - N) runs of system 1 are generated with antithetics. Hence Table 6 represents alternative B.

C. Joint Application of Antithetic Variates and Common Random Numbers.

This alternative has already been shown in Table 4. The tables 4, 5 and 6 need refinement if we take into account uneven values of M and N as we shall see in Table 7 later on.

Next we shall derive the variance of the estimated difference $\bar{d}$ for each of the three alternatives. The derivation of this variance is based on (249) above. We first discuss the following symbols.

$$c_1 = \text{cov}(\underline{x}_i, \underline{x}_g) \quad \text{if } \underline{x}_i \text{ and } \underline{x}_g \text{ are generated with negative correlation} \tag{255}$$

$$c_2 = \text{cov}(\underline{y}_j, \underline{y}_h) \quad \text{if } \underline{y}_j \text{ and } \underline{y}_h \text{ are generated with negative correlation} \tag{256}$$

$$c_3 = \text{cov}(\underline{x}_i, \underline{y}_j) \quad \text{if } \underline{x}_i \text{ and } \underline{y}_j \text{ are generated with positive (cross-) correlation} \tag{257}$$

$$c_4 = \text{cov}(\underline{x}_i, \underline{y}_j) \quad \text{if } \underline{x}_i \text{ and } \underline{y}_j \text{ are generated with negative (cross-) correlation} \tag{258}$$

Regarding the symbol c_1 we observe that if $\underline{x}_1$ and $\underline{x}_2$ are generated with antithetic variates and so are $\underline{x}_3$ and $\underline{x}_4$, etc., then their covariances are equal, i.e.,

$$\text{cov}(\underline{x}_1, \underline{x}_2) = \text{cov}(\underline{x}_3, \underline{x}_4) = \cdots = \text{cov}(\underline{x}_{M-1}, \underline{x}_M) = c_1 \tag{259}$$

where M is supposed to be even. It is easy to see that (259) holds since

$$\underline{x}_1 = g(\vec{\underline{R}}_1) \quad \text{and} \quad \underline{x}_2 = g(\vec{I} - \vec{\underline{R}}_1) \tag{260}$$

$$\underline{x}_3 = g(\vec{\underline{R}}_2) \quad \text{and} \quad \underline{x}_4 = g(\vec{I} - \vec{\underline{R}}_2) \tag{261}$$

where the function g denotes the computer program that makes the response a function of the random numbers; $\vec{\underline{R}}_1$ and $\vec{\underline{R}}_2$ have the same distribution. In the same way we find that if $\underline{y}_1$ and $\underline{y}_2$, $\underline{y}_3$ and $\underline{y}_4$, etc. are generated from antithetic variates, then (262) holds.

$$\text{cov}(\underline{y}_1, \underline{y}_2) = \text{cov}(\underline{y}_3, \underline{y}_4) = \cdots = \text{cov}(\underline{y}_{N-1}, \underline{y}_N) = c_2 \tag{262}$$

where N is taken even. Further, if run j $(j = 1, 2, \ldots, N)$ of both systems is generated from the same random numbers, then (263) holds.

$$\text{cov}(\underline{x}_1, \underline{y}_1) = \text{cov}(\underline{x}_2, \underline{y}_2) = \cdots = \text{cov}(\underline{x}_N, \underline{y}_N) = c_3 \tag{263}$$

Finally, if both antithetics and common random numbers are applied then undesirable negative cross-correlations are created, i.e.,

$$\text{cov}(\underline{x}_1, \underline{y}_2) = \text{cov}(\underline{x}_2, \underline{y}_1) = \cdots = \text{cov}(\underline{x}_N, \underline{y}_{N-1}) = c_4 \tag{264}$$

Note that the possibility of generating correlated runs depends on the values of M and N in the following two ways.

(i) If the number of runs for a system is odd, then one run cannot be generated antithetically.

(ii) As we saw in the discussion of alternative B, if the number of runs for both systems is different $(M > N)$, then we can match only the first N runs of system 1 with the runs of system 2 (taking the same random numbers for a run with system 1 and 2).

We illustrate the derivation of $\text{var}(\bar{\underline{d}})$ considering the situation in Table 7 where only antithetics are applied (corresponding with Table 5) and where $M = N =$ even. The first term in the expression for $\text{var}(\bar{\underline{d}})$ in (249) reduces to

$$
\begin{aligned}
&M^{-2}\{\text{cov}(\underline{x}_1, \underline{x}_2) + \text{cov}(\underline{x}_2, \underline{x}_1) + \text{cov}(\underline{x}_3, \underline{x}_4) + \text{cov}(\underline{x}_4, \underline{x}_3) \\
&\quad + \cdots + \text{cov}(\underline{x}_{M-1}, \underline{x}_M) + \text{cov}(\underline{x}_M, \underline{x}_{M-1})\} \\
&= M^{-2}\{2\ \text{cov}(\underline{x}_1, \underline{x}_2) + 2\ \text{cov}(\underline{x}_3, \underline{x}_4) + \cdots + 2\ \text{cov}(\underline{x}_{M-1}, \underline{x}_M)\} \\
&= M^{-2}\{2\ \frac{M}{2}\ c_1\} = M^{-1}\ c_1 \qquad (265)
\end{aligned}
$$

In the same way we find that the third term (249) reduces to $N^{-1}\ c_2$. Since no correlation exists between runs of different systems the last term in (249) vanishes. Hence

$$\text{var}(\bar{\underline{d}}) = M^{-1}\ c_1 + M^{-1}\ \sigma_1^2 + N^{-1}\ c_2 + N^{-1}\ \sigma_2^2 \qquad (266)$$

The derivation of $\text{var}(\bar{\underline{d}})$ for the other situations is analogous. Because (249) always contains the two terms $(M^{-1}\ \sigma_1^2 + N^{-1}\ \sigma_2^2)$ these terms are not shown in Table 7.[50)]

III.8.4. Comparisons Among Alternatives

From Table 7 it follows that in order to determine which of the three alternatives A, B, C, gives the lowest variance, we need to know the <u>relative magnitudes of the covariances</u> c_1 through c_4. For instance, if $(c_3 + c_4) > 0$ or equivalently $c_3 > |c_4|$ then method C is better than A; method A is better than B if $c_3 < |c_1 + c_2|/2$ (and $M = N =$ even); method B is better than C if $|c_4| > |c_1 + c_2|/2$ (and $M = N =$ even), etc. We like to know if such relations among the covariances hold <u>in general</u> in the simulation of systems. Therefore let us first consider some very <u>simple</u> "systems" and let us derive <u>analytically</u>, whether such relations can be assumed to hold for all systems. In the "systems" under consideration, the response $\underline{x}$ of a run depends on a single random number $\underline{r}$ and the response is a monotonic increasing function of $\underline{r}$. For instance

TABLE 7. The Variance of the Estimated Difference between the Responses of Systems 1 and 2; to Each Entry the Common Terms $(M^{-1}\sigma_1^2 + N^{-1}\sigma_2^2)$ Should Be Added

Case		A: antithetic variates only	B: common random numbers in the first N runs (antithetics in the last M - N runs of system 1)
I	M = N even	$M^{-1}c_1 + N^{-1}c_1$	$-2M^{-1}c_3$
II	M = N = odd	$(M^{-1} - M^{-2})c_1^2 + (N^{-1} - N^{-2})c_2$	$-2M^{-1}c_3$
III	M(even) > N(even)	$M^{-1}c_1 + N^{-1}c_2$	$-2M^{-1}c_3 + M^{-2}(M - N)c_1$
IV	M(even) > N(odd)	$M^{-1}c_1 + (N^{-1} - N^{-2})c_2$	$-2M^{-1}c_3 + M^{-2}(M - N - 1)c_1$
V	M(odd) > N(even)	$(M^{-1} - M^{-2})c_1 + N^{-1}c_2$	$-2M^{-1}c_3 + M^{-2}(M - N - 1)c_1$
VI	M(odd) > N(odd)	$(M^{-1} - M^{-2})c_1 + (N^{-1} - N^{-2})c_2$	$-2M^{-1}c_3 + M^{-2}(M - N)c_1$

Case		C: joint application of antithetic variates and common random numbers in all runs
I	M = N = even	$M^{-1}c_1 + N^{-1}c_2 - 2M^{-1}(c_3 + c_4)$
II	M = N = odd	$(M^{-1} - M^{-2})c_1 + (N^{-1} - N^{-2})c_2 - 2M^{-1}\{c_3 + (1 - M^{-1})c_4\}$
III	M(even) > N(even)	$M^{-1}c_1 + N^{-1}c_2 - 2M^{-1}(c_3 + c_4)$
IV	M(even) > N(odd)	$M^{-1}c_1 + (N^{-1} - N^{-2})c_2 - 2M^{-1}(c_3 + c_4)$
V	M(odd) > N(even)	$(M^{-1} - M^{-2})c_1 + N^{-1}c_2 - 2M^{-1}(c_3 + c_4)$
VI	M(odd) > N(odd)	$(M^{-1} - M^{-2})c_1 + (N^{-1} - N^{-2})c_2 - 2M^{-1}(c_3 + c_4)$

$$\underline{x} = \underline{r}^2 \qquad (267)$$

Hence the antithetic run is

$$\underline{x}_a = (1 - \underline{r})^2 \qquad (268)$$

So

$$\begin{aligned} c_1 &= \text{cov}(\underline{x}, \underline{x}_a) = E(\underline{x}\ \underline{x}_a) - E(\underline{x})\ E(\underline{x}_a) \\ &= E[\underline{r}^2(1 - \underline{r})^2] - E(\underline{r}^2)\ E[(1 - \underline{r})^2] \\ &= E(\underline{r}^2 - 2\underline{r}^3 + \underline{r}^4) - E(\underline{r}^2)\ E(1 - 2\underline{r} + \underline{r}^2) \\ &= \int_0^1 (r^2 - 2r^3 + r^4)\ dr - \int_0^1 r^2\ dr \int_0^1 (1 - 2r + r^2)\ dr \\ &= 1/30 - (1/3)(1/3) = -7/90 \end{aligned} \qquad (269)$$

Other simple "systems" are specified in the columns (1) and (2) of Table 8; the resulting covariances c_1 through c_4 have been calculated analogous to (269) and are given in the columns (3) through (6) of Table 8; the columns (7) through (9) show some comparisons among covariances, these comparisons being suggested above. The comparisons in Table 8 prove by counterexample that no relation among the covariances holds for all systems, i.e., none of the methods A, B, or C is best in all circumstances.

Besides the analytic results for the simple systems of Table 8 we obtained experimental results for some more <u>complicated</u> <u>systems</u> (see Table 9):

(1) The <u>single</u>-server queuing system was simulated for the following two versions.

(a) System 1 has exponentially distributed interarrival and service times with parameters 1.5 and 2, respectively; system 2 also has exponential arrival and service times with parameters 0.5 and 2.5, respectively.

(b) System 1 has exponential arrival and service times with parameters 1.5 and 2; system 2 has exponential arrival times with parameter 1.5 and constant service times with value 1/2.

TABLE 8. The Relative Magnitudes of the Covariances in Some Simple Systems

Systems		Covariances				Comparisons		
$\underline{x} = x(\underline{r})$	$\underline{y} = y(\underline{r})$	c_1	c_2	c_3	c_4	$\frac{\lvert c_1 + c_2 \rvert}{2c_3}$	$\frac{\lvert c_1 + c_2 \rvert}{2\lvert c_4 \rvert}$	$\frac{c_3}{\lvert c_4 \rvert}$
(1)	(2)	(3)	(4)	(5)	(6)	(7)	(8)	(9)
r^2	$r^2/2$	-7/90	-7/360	8/810	-7/180	> 1	> 1	> 1
$2r + 5$	$r^2 + r$	-1/3	-59/180	1/3	-1/3	< 1	< 1	.
$2r^2$	$\sqrt{r}$	.	.	8/63	-132/905	.	.	< 1

(2) The queuing system with four service stations in sequence was simulated for two variants, the two variants differing in the parameters of the four exponential service time distributions. The system is described in more detail in Naylor et al.[51] (1967b, pp. 703-705).

All single-server systems were replicated 50 times, i.e., M = N = 50; the four-stations systems were run twenty times so M = N = 20. These experiments were performed applying the methods A, B and C and moreover "method D," the crude method were all runs are independent. The estimated variances of the estimated difference between the responses of two systems when applying the methods A through D are given in Table 9.[52] This table suggests that the smallest variance is not necessarily obtained when we apply both antithetic variates and common random numbers (method C). Actually for the single-server systems we found the order C, A, B, D but for the four-stations system the order B, C, A, D. (Obviously we are neglecting possible sampling errors.)[53]

TABLE 9. The Estimated Variance of the Estimated Difference between the Responses of Two Systems

System	Method			
	A (antithetics)	B (common)	C (joint)	D (independent)
1. Single-server				
a. Exponential	0.057	0.11	0.053	0.12
b. Exp/Constant	0.090	0.18	0.081	0.33
2. Four-stations	0.022	0.0019	0.0032	0.037

III.8.5. Optimum Alternative and Computer Time Allocation

Both the analytical results for the extremely simple systems and the experimental results for the more complex systems showed that no alternative is best for all systems. Therefore we shall assume that it is decided to generate some pilot-runs for the two systems and to estimate the variances σ_1^2 and σ_2^2 and the covariances c_1 through c_4 from these pilot-runs in order to select the method that will be applied in the remaining runs. (In Section III.8.6 we shall consider the estimation procedure for the variances and covariances in detail.) If we want to use these estimated variances and covariances in Table 7, then we need to know which of the six cases listed in that table will be valid. So we need to know if M and N are even or odd, and if $M = N$, $M > N$ (or $M < N$, for in Table 7 we called the system with the smallest number of runs "system 2" but if we do not know beforehand that system 2, associated with σ_2^2 and c_2, will be run less often than system 1 then we have also to consider the case $M < N$). Therefore we shall fix M and N after the pilot-phase in such a way that the available computer time is consumed, i.e., (270) should be satisfied with the additional restrictions (271) and (272).

$$t_1M + t_2N = T \tag{270}$$

$$M \geq N_p \tag{271}$$

$$N \geq N_p \tag{272}$$

where N_p is the number of pilot-runs obtained for both system 1 and 2; t_1 and t_2 are the computer time per run of systems 1 and 2, respectively; T is the total computer time available for simulating the two systems; $(M - N_p)$ and $(N - N_p)$ are the number of runs to be generated after the pilot-phase. A unique solution can be calculated if we add the restriction that we simulate both systems an equal number of times. (We shall return to this restriction in a moment.) So

$$M = N \tag{273}$$

Given the M and N resulting from (270) through (273), and given the estimated variances and covariances we are able to use Table 7 and find the method that is expected to yield the lowest variance. We point out that the introduction of a limited amount of computer time in (270) is necessary since for an unlimited amount of computer time M and N could be taken infinitely large. M and N infinite would mean that $\text{var}(\bar{d})$ is zero for all three methods A, B, C and consequently the choice of a particular method would be indifferent.

Of course there is no need to take M and N equal as in (273). Instead we may choose M and N in such a way that $\text{var}(\bar{d})$ is minimized given the computer time restriction in (270) and the side-conditions (271) and (272). The fact that the formula for $\text{var}(\bar{d})$ varies with M and N and with the methods A, B, C complicates the optimal choice of M and N. For besides the six cases for M and N listed in Table 7 we have to add four more cases where $M < N$ (since as we remarked above, system 2 associated with σ_2^2, c_2 and N may be run more often than system 1). For the sake of simplicity we decided to use an approximate formula for $var(\bar{d})$ based on dropping term in M^{-2} and N^{-2} in Table 7. This results in Table 10.[54) From Table 10 it follows that $var(\bar{d})$ can be approximated by

$$var(\bar{d}) = a_1 M^{-1} + a_2 N^{-1} \quad (274)$$

the values of the coefficients a_1 and a_2 varying with the methods A, B, C and with the cases $M \geq N$ or $M \leq N$. Summarizing, we want to minimize the variance of the estimated difference between the responses of the two systems specified in (274), given the restriction of a limited amount of computer time specified in (270) and given the condition that we have already taken N_p pilot-runs of system 1 and 2 as shown by (271) and (272).

The optimum values of M and N in (274) depend on the signs of the coefficients a_1 and a_2. In Appendix 2 we prove that we have the following two possibilities:

(1) Both coefficients a_1 and a_2 are positive.

(2) One coefficient is positive and the other coefficient is negative.

TABLE 10. Approximation for $var(\bar{d})$, Based on Dropping Terms in M^{-2} and N^{-2}

Case	Method	
$M \geq N$	A	$(\sigma_1^2 + c_1)M^{-1} + (\sigma_2^2 + c_2) N^{-1}$
	B	$(\sigma_1^2 - 2c_3)M^{-1} + \sigma_2^2 N^{-1}$
	C	$(\sigma_1^2 + c_1 - 2c_3 - 2c_4)M^{-1} + (\sigma_2^2 + c_2) N^{-1}$
$M \leq N$	A	$(\sigma_1^2 + c_1)M^{-1} + (\sigma_2^2 + c_2) N^{-1}$
	B	$(\sigma_1^2 M^{-1} + (\sigma_2^2 - 2c_3) N^{-1}$
	C	$(\sigma_1^2 + c_1)M^{-1} + (\sigma_2^2 + c_2 - 2c_3 - 2c_4) N^{-1}$

In Appendix 13 we derive that if both coefficients are positive, then $\text{var}(\bar{\underline{d}})$ in (274) is minimized subject to the computer time restriction (270) if we choose M and N equal to (275) and (276), respectively.

$$M^* = T/\{t_1 + (a_1^{-1} a_2 t_1 t_2)^{1/2}\} \qquad (275)$$

$$(a_1, a_2 > 0)$$

$$N^* = T/\{t_2 + (a_1 a_2^{-1} t_1 t_2)^{1/2}\} \qquad (276)$$

From Fig. 8 we see that the solutions in (275) and (276) may violate the conditions (271) and (272) requiring that M and N have at least the value N_p. Moreover, if we use a particular value for a_1 and a_2 in (274) then these values imply that either $M \leq N$ or $M \geq N$ as Table 10 showed. Because of these additional restrictions the optimal values of M and N should be determined from Table 11 as we prove in Appendix 13. Table 11 shows that if (276) yields $N^* < N_p$ then N_0, the optimal value of N, is set equal to the pilot-number N_p; if (276) yields $N^* > T/(t_1 + t_2)$ so the condition $M \geq N$ is violated, then we take an equal number of runs, i.e., $M_0 = N_0 = T/(t_1 + t_2)$.

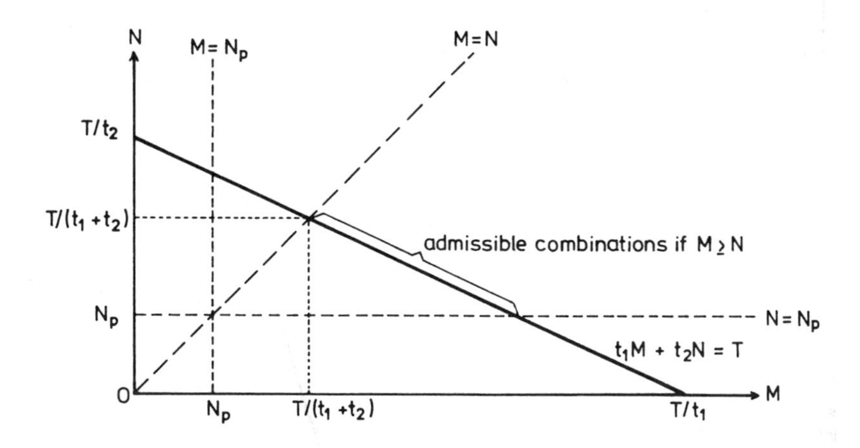

Fig. 8. Admissible combinations of M and N.

TABLE 11. Optimal Values of M and N if a_1 and a_2 Are Positive (with M^* and N^* Specified in (275) and (276), respectively)

Case	Side Conditions	M_0	N_0
$M \geq N$	$N_p \leq N^* \leq T/(t_1 + t_2)$	M^*	N^*
	$N^* < N_p$	$(T - t_2N_p)/t_1$	N_p
	$N^* > T/(t_1 + t_2)$	$T/(t_1 + t_2)$	$T/(t_1 + t_2)$
$M \leq N$	$N_p \leq M^* \leq T/(t_1 + t_2)$	M^*	N^*
	$M^* < N_p$	N_p	$(T - t_1N_p)/2$
	$M^* > T/(t_1 + t_2)$	$T/(t_1 + t_2)$	$T/(t_1 + t_2)$

If one of the coefficients a_1 and a_2 is <u>negative</u> then, as we derive in Appendix 13, we should take M and N as in (277).

$$M_0 = N_0 = T/(t_1 + t_2) \quad (a_1 \text{ or } a_2 \text{ negative}) \tag{277}$$

There is a restriction not mentioned until now, namely, M and N must be <u>integer</u>. Hence M_0 and N_0 determined from Table 11 or (277) should be made integer. We can check the integer points in the neighborhood of M_0 and N_0 as Fig. 9 demonstrates. For the sake of simplicity we may decide to restrict our attention to the following three pairs where square brackets denote the integer part.

$$M = [M_0], \qquad N = [N_0] \tag{278}$$

$$M = [M_0] + 1, \qquad N = [N_0] \tag{279}$$

$$M = [M_0] \qquad N = [N_0] + 1 \tag{280}$$

We have to check if the integer values still satisfy the computer time and the pilot-run conditions (270) through (272), and either $M \geq N$ or $M \leq N$. (Only the computer time restriction (270) is shown in Fig. 9.) If more than one pair of integer values of M and N satisfy all these conditions then we select the pair that yields the smallest variance. Using (278) through (280) we find

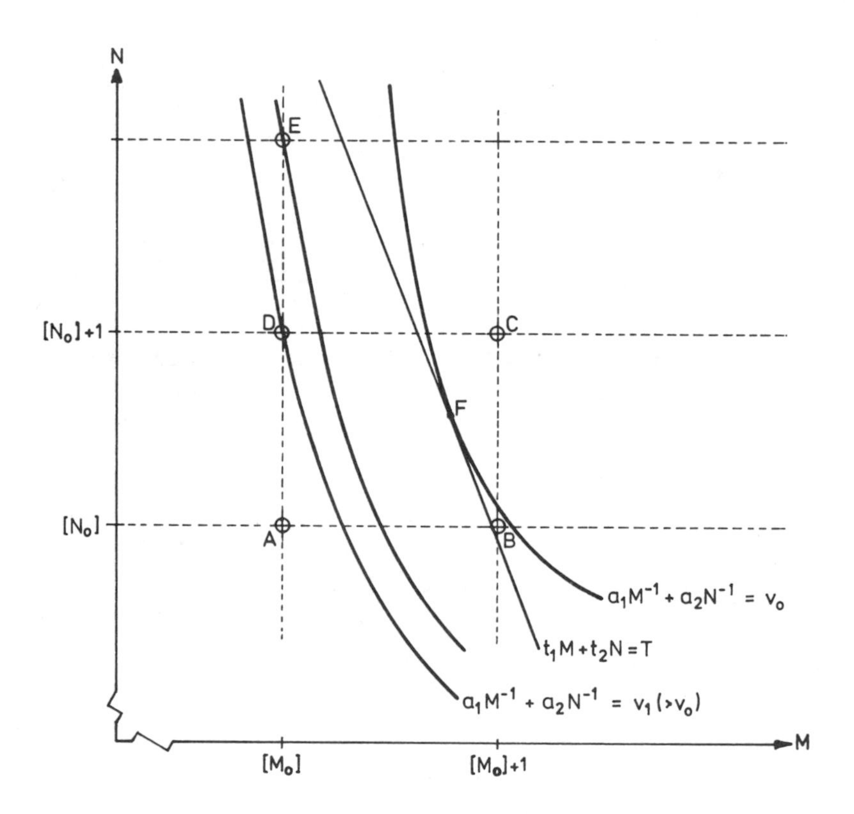

Fig. 9. The optimum combination of the integer values of M and N. F: optimum (M_0, N_0); A, B, C, D, E: integer combinations; C, B: inadmissible (see computer time restriction); E better than D; E not listed in (278) through (280).

the pair of integer values of M and N that results in the smallest variance. However, this minimum variance is derived for a particular pair of values of the coefficients a_1 and a_2. From Table 10 it follows that a different pair of coefficients holds for each method and for each case ($M \geq N$ or $M \leqslant N$); so six pairs (a_1, a_2) result. Actually method A gives the same coefficients for the cases $M \geq N$ and $M \leq N$ so there are five instead of six different pairs of values of a_1 and a_2. Each of the five pairs (a_1, a_2) gives a corresponding pair of optimum values for M and N. Substitution of the optimum M and N into the expression for $\text{var}(\bar{d})$ in (274) yields the minimum variance. In this way five minimum variances will be found. Finally we select the minimum among these five minimal variances and determine the corresponding method (A, B, or C) and number of runs, (M and N).

Note that in our selection procedure we have taken into account the covariances created by the various methods. For these covariances determine the values of the coefficients a_1 and a_2 in (274). Moreover we can incorporate possible differences among the computer times per run per method. For we can take values for t_1 and t_2 in the computer time restriction (270) that differ with the methods A, B, and C. Observe that the variance is minimized not only by selecting an appropriate variance reducing method but also by choosing an optimal combination of the number of runs per system. Once we have decided on the optimal number of runs of each system we may get a more accurate estimate of $\text{var}(\bar{d})$ by using the exact formula for $\text{var}(\bar{d})$ based on Table 7.

III.8.6. Estimation of the Coefficients of the Optimization Procedure

In this section we shall discuss how the pilot-runs can be used to obtain estimates of the coefficients a_1 and a_2 in (274) and t_1 and t_2 in (270). The N_p pilot-runs of system 1 and 2 should be generated using method C, joint application of antithetic variates and common random numbers. For method C generates all four covariances c_1 through c_4. We can estimate c_1, the covariance between antithetic runs of system 1, by

$$\underline{\hat{c}}_1 = \sum_{i=1}^{n_p} (\underline{x}_{2i-1} - \bar{\underline{x}})(\underline{x}_{2i} - \bar{\underline{x}}_a)/(n_p - 1) \tag{281}$$

where

$$\bar{\underline{x}} = \sum_{i=1}^{n_p} \underline{x}_{2i-1}/n_p \tag{282}$$

$$\bar{\underline{x}}_a = \sum_{1}^{n_p} \underline{x}_{2i}/n_p \tag{283}$$

$$n_p = N_p/2 \quad (N_p \text{ is even}) \tag{284}$$

The estimation of c_2 is completely analogous to (281). When estimating c_3, the positive cross-covariance between $\underline{x}_i$ and $\underline{y}_i$ ($i = 1, \ldots, N_p$), we have to take into account that the well-known estimation formulas like (281) are based on the assumption that each pair of observations is <u>independent</u> of each other pair. However, the pair $(\underline{x}_1, \underline{y}_1)$ is not independent of the pair $(\underline{x}_2, \underline{y}_2)$ since both pairs use the random numbers $\vec{\underline{R}}_1$ as Table 4 showed. We can solve this complication by dividing the N_p pilot-runs into two groups as in Table 12. Hence the positive cross-covariance can be estimated from

$$\underline{\hat{c}}_3(1) = \sum_{i=1}^{n_p} (\underline{x}_{2i-1} - \bar{\underline{x}})(\underline{y}_{2i-1} - \bar{\underline{y}})/(n_p - 1) \tag{285}$$

$$\underline{\hat{c}}_3(2) = \sum_{i=1}^{n_p} (\underline{x}_{2i} - \bar{\underline{x}}_a)(\underline{y}_{2i} - \bar{\underline{y}}_a)/(n_p - 1) \tag{286}$$

$$\underline{\hat{c}}_3 = \{\underline{\hat{c}}_3(1) + \underline{\hat{c}}_3(2)\}/2 \tag{287}$$

In the same way we can estimate c_4 grouping the observations as in Table 13.[55] The well-known formula for the estimation of the variance is also based on independent observations. Hence we can

TABLE 12. Grouping of N_p Pilot-Runs for the Estimation of the Covariance c_3

Group 1			Group 2		
Random numbers	Response of system 1	Response of system 2	Random numbers	Response of system 1	Response of system 2
$\vec{R}_1$	x_1	y_1	$\vec{I} - \vec{R}_1$	x_2	y_2
$\vec{R}_2$	x_3	y_3	$\vec{I} - \vec{R}_2$	x_4	y_4
$\vdots$	$\vdots$	$\vdots$	$\vdots$	$\vdots$	$\vdots$
$\vec{R}_{n_p}$	x_{N_p-1}	y_{N_p-1}	$\vec{I} - \vec{R}_{n_p}$	x_{N_p}	y_{N_p}

TABLE 13. Grouping the N_p Pilot-Runs for the Estimation of c_4

Group 1				Group 2			
System 1		System 2		System 1		System 2	
Random numbers	Response	Random numbers	Response	Random numbers	Response	Random numbers	Response
$\vec{R}_1$	x_1	$\vec{I} - \vec{R}_1$	y_2	$\vec{I} - \vec{R}_1$	x_2	$\vec{R}_1$	y_1
$\vec{R}_2$	x_3	$\vec{I} - \vec{R}_2$	y_4	$\vec{I} - \vec{R}_2$	x_4	$\vec{R}_2$	y_3
$\vdots$	$\vdots$	$\vdots$	$\vdots$	$\vdots$	$\vdots$	$\vdots$	$\vdots$
$\vec{R}_{n_p}$	x_{N_p-1}	$\vec{I} - \vec{R}_{n_p}$	y_{N_p}	$\vec{I} - \vec{R}_{n_p}$	x_{N_p}	$\vec{R}_{n_p}$	y_{N_p-1}

use the grouping in Tables 12 or 13 to estimate the variance of system 1 from

$$\hat{\underline{\sigma}}_1^2(1) = \sum_{i=1}^{n_p} (\underline{x}_{2i-1} - \bar{\underline{x}})^2/(n_p - 1) \tag{288}$$

$$\hat{\underline{\sigma}}_1^2(2) = \sum_{i=1}^{n_p} (\underline{x}_{2i} - \bar{\underline{x}}_a)^2/(n_p - 1) \tag{289}$$

$$\hat{\underline{\sigma}}_1^2 = \{\hat{\underline{\sigma}}_1^2(1) + \hat{\underline{\sigma}}_1^2(2)\}/2 \tag{290}$$

The same procedure can be applied for the estimation of σ_2^2. The estimators for the variances and covariances can be substituted into Table 10 in order to obtain the estimators for the coefficients a_1 and a_2 in the expression for $\text{var}(\bar{\underline{d}})$ in (274). Note that the estimates of the variances and covariances are obtained by applying only method C; yet these estimates can be used in Table 10 to obtain estimates of a_1 and a_2 for A, B and C (and for $M \geq N$, $M \leq N$).

If we need $\underline{T}_p$ units of time to run system 1, N_p times applying method C, then we can estimate t_1 in the computer time restriction (270) by

$$\hat{\underline{t}}_1 = \underline{T}_p/N_p \tag{291}$$

In the same way we can estimate the running time for system 2 when applying C.[56] As we mentioned in Sections III.6 and III.7 the extra computer time for the application of antithetics or common random numbers is usually negligible. Hence we may decide to use the same $\hat{\underline{t}}_1$ and $\hat{\underline{t}}_2$ for all three methods A, B, and C. However, if we generate antithetic runs by subtracting $\underline{r}$ from the number one, then we can estimate the time per run of system 1 when not using antithetics by keeping track of the time needed to generate $x_1(\vec{R}_1)$, $x_3(\vec{R}_2)$, ... , $x_{N_p-1}(\vec{R}_{n_p})$ and dividing this total time by $N_p/2$.

We observe that the optimum values of M, N and $\mathrm{var}(\bar{d})$ are nonlinear functions of the variances, covariances and times per run. Hence, when using unbiased estimators for these variances, etc., the corresponding estimated optimum M, N, and $\mathrm{var}(\bar{d})$ are still biased.[57)] This type of bias is mentioned by Fishman (1967, pp. 19-22). We have not tried to determine this bias in our problem but assume that it is of negligible importance.[58)]

Summarizing, we obtain the following procedure for the selection of one of the methods A, B, and C and for the determination of the optimal combination of runs of system 1 and 2.

(1) Take N_p pilot-runs for system 1 and N_p pilot-runs for system 2, applying method C (joint application of antithetics and common random numbers as specified in Table 4).

(2) Estimate the variances σ_1^2 and σ_2^2 defined in (250) and (251), the covariances c_1 through c_4, defined in (255) through (258), and the time per run of system 1 and 2, i.e., t_1 and t_2 in (270). The estimation formulas are given in (281) through (291).

(3) Substitute the estimated variances and covariances into Table 10 to obtain estimates for the coefficients a_1 and a_2 of $\mathrm{var}(\bar{d})$ in (274). There are six pairs of estimated coefficients a_1 and a_2 since we have three methods (A, B, C) and two cases ($M \geq N$, $M \leq N$); five of these six pairs are different.

(4) For each of the five pairs of estimated a_1 and a_2, and the corresponding estimated t_1 and t_2, determine the optimum combination of M and N (the number of runs for system 1 and 2) using Table 11 and formulas (277) through (280).

(5) For each of the five pairs of optimum values of M and N determine the corresponding variance using the approximation for $\mathrm{var}(\bar{d})$ in (274) or the exact formulas in Table 7.

(6) Determine the minimum of the five variances of step 5 and the corresponding method and number of runs.

III.8.7. Comments on the Optimization Procedure

Let us first consider some applications of the optimization procedure. The procedure was applied to several simple queuing systems. The results are summarized in Table 14.[59] Application of the procedure to the single-server systems 1a and 1b of Table 9 yielded the same ranking as in Table 9, i.e., in order of increasing $\hat{var}(\bar{d})$ the ranking is C, A, B, D. Application to four other situations where systems with one or two service stations were compared, resulted in the ranking C, B, A, D. The optimum values of M and N were found to differ greatly from each other (unless there is a negative coefficient a_1 or $N^* > T/(t_1 + t_2)$; then $M_0 = N_0$).

We observe that it is possible that the ordering resulting from the optimization procedure differs from the ordering found when applying each of the four methods to the two systems an equal number of times, say N_p times (the latter procedure was applied in Table 9). This is illustrated in Fig. 10, where A_p and B_p denote the variance of $\bar{d}$ when applying the methods A and B, N_p times each; A_0 and B_0 denote the minimum variance when applying the methods A and B satisfying the computer time restriction. In Fig. 10 we have $A_p < B_p$ but $A_0 > B_0$. (Further, a different ranking may be explained by the fact that the optimization procedure gives only an estimate of the variance and is based on the approximation for $var(\bar{d})$ in (274) above.)

We realize that our optimization procedure has several drawbacks.

(i) It takes some time to estimate the coefficients and to perform the necessary calculations with these coefficients. Nevertheless in a complicated simulation study this extra time is negligible and therefore the optimization procedure may be worthwhile.

(ii) The procedure is based on estimates of the variances and covariances calculated from the N_p pilot-runs of each system. Hence if N_p increases then these estimates become more reliable and a more reliable selection from the methods A, B, C is possible.

TABLE 14. Results of the Application of the Optimization Procedure

Systems compared	Minimal var($\underline{\bar{d}}$) ($\times 10^5$)				Optimum M and N for method C*)		Comment on M and N
	A	B	C	D	M	N	
1. Single-server systems							
a. Exponential distributions+)	162	76	52	257	100	100	$N^* > T/(t_1 + t_2)$
b. Exponential distributions+)	83	99	77	104	139	50	$N^* < N_p$
c. Constant service times	9	3	2	14	150	150	a_1 negative
d. One Exponential and one constant service time	114	145	113	180	164	54	
2. One single-server and one two-servers system	33	26	24	51	128	68	
3. Two two-servers systems	61	45	39	79	95	94	a_1 negative

*) For the methods A, B and D about the same values were found.

+) 1a and 1b use different computer programs.

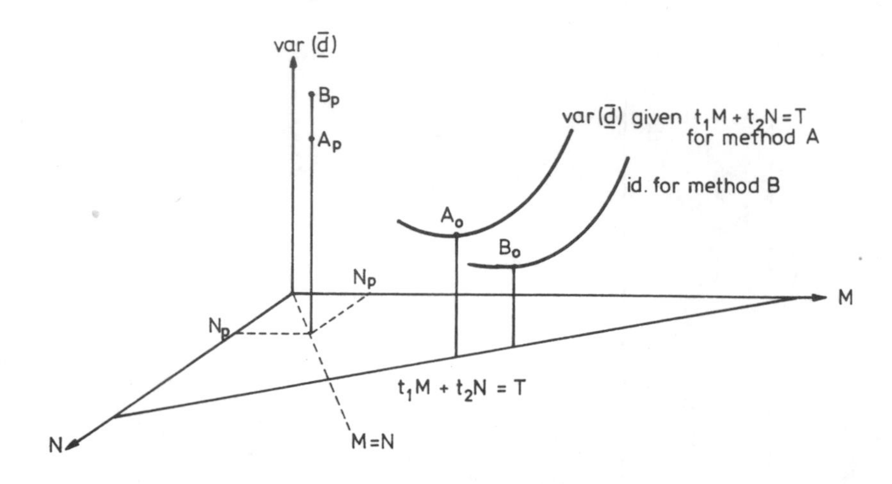

Fig. 10. The variance of $\bar{d}$ when the number of runs is the same ($M = N = N_p$) or when the number of runs is optimal given the computer time restriction

Unfortunately if we augment N_p and we find that the best method is not C (the method applied in the pilot-phase) but either A or B, then most of the runs have already been generated with the inferior method C. To solve this dilemma we may decide to restrict N_p to, say, 10% of our a priori guess of M_0 and N_0. (Actually a more accurate procedure for the minimal var($\bar{d}$) would not compare var($\bar{d}$) when applying A, B, or C in all M_0 and N_0 runs but would compare var($\bar{d}$) when applying A, B or C in $(M_0 - N_p)$ and $(N_0 - N_p)$ runs and C in N_p runs. When N_p is small compared with M_0 and N_0--as it should be--then for the sake of simplicity we may use the procedure as summarized at the end of Section III.8.6.)

(iii) If we decide after the pilot-phase to switch from method C to either method A or B, then we get "<u>nonhomogeneous</u>" output. Switching from C to A means that the cross-covariances c_3 and c_4 become zero; switching from C to B means that we get no more antithetic runs and c_1 and c_2 become zero. This complicates the statistical analysis. For, if we used C in all M and N runs then we could estimate the variance of $\bar{d}$ using the formula for var($\bar{d}$) in Table 7 and estimation formulas analogous to (281) through (290). If we switch from C to A after the pilot-phase then we can use

$$\mathrm{var}(\bar{d}) = \mathrm{var}(\bar{x}) + \mathrm{var}(\bar{y}) - 2\,\mathrm{cov}(\bar{x}, \bar{y}) \tag{292}$$

For even if we switch from C to A we still have M/2 antithetic pairs.[60] Hence var($\bar{x}$) can be estimated by taking the average of each antithetic pair as in (239) through (241). In the same way we estimate var($\bar{y}$) from N/2 antithetic pairs. The last term in (292) can be estimated from

$$\mathrm{cov}(\bar{x}, \bar{y}) = \sum_{i=1}^{M} \sum_{j=1}^{N} \mathrm{cov}(x_i, y_j)/(MN)$$

$$= N_p(c_3 + c_4)/(MN) \tag{293}$$

where the last equality in (293) holds since x_i and y_j are correlated in the first N_p pilot-runs of each system but are independent after the pilot-phase. If after the pilot-phase we switch from C to B then the last term in (292) can be evaluated from

$$\mathrm{cov}(\bar{x}, \bar{y}) = \{\min(N, M)c_3 + N_p c_4\}/(MN) \tag{294}$$

When switching from C to B the variance of $\bar{x}$ can be estimated by dividing the M observations x_i into two parts, part 1 consisting of antithetic pairs and part 2 of independent observations x. Then $\bar{x}$ is a weighted average of the averages of these two parts as in (295).

$$\bar{x} = a\bar{x}_1 + (1 - a)\,\bar{x}_2 \tag{295}$$

So

$$\mathrm{var}(\bar{x}) = a^2\,\mathrm{var}(\bar{x}_1) + (1 - a)^2\,\mathrm{var}(\bar{x}_2) \tag{296}$$

where $var(\bar{\underline{x}}_1)$ can be estimated analogous to (239) through (241) and $var(\bar{\underline{x}}_2)$ can be estimated in the traditional way. Once we have estimated $var(\bar{\underline{d}})$ we may determine confidence limits for the difference $E(\bar{\underline{d}})$ using the normal distribution, at least if we have many observations. In that case we have

$$\bar{\underline{d}} = \bar{\underline{x}} - \bar{\underline{y}} \tag{297}$$

with $\bar{\underline{x}}$ and $\bar{\underline{y}}$ being asymptotically normal as follows from the central limit theorem.[61] The normality of $\bar{\underline{x}}$ and $\bar{\underline{y}}$ can also be based on the asymptotic normality of the individual observations $\underline{x}_i$ and $\underline{y}_j$ if the "stationary m-dependent central limit theorem" holds for these individual observations; this theorem was mentioned in the discussion of (12) and (13) in Section III.2.1. So we assume that $\bar{\underline{x}}$ and $\bar{\underline{y}}$ are asymptotically bivariate normal.[62] Hence $\bar{\underline{d}}$, a linear transformation of this two-dimensional normal variable, is also asymptotically normally distributed as follows from Fisz (1967, p. 162). So if we have many observations we can use the $(1 - \alpha)$ confidence interval

$$\bar{\underline{d}} \pm z^{\alpha/2}\underline{s} \tag{298}$$

where $z^{\alpha/2}$ is the upper $\alpha/2$ point of the standard normal variable and $\underline{s}$ is the estimator of the standard deviation of $\bar{\underline{d}}$ calculated from (292). If we have only a limited number of observations then (298) should be replaced by an interval based on the t-distribution:

$$\bar{\underline{d}} \pm t_{n-1}^{\alpha/2}\, \underline{s}_d / \sqrt{n} \tag{299}$$

where

$$\underline{s}_d = \left\{ \sum_{i=1}^{n} (\underline{d}_i - \bar{\underline{d}})^2/(n - 1) \right\}^{1/2} \tag{300}$$

$$\bar{\underline{d}} = \sum_{i=1}^{n} \underline{d}_i/n \tag{301}$$

and each $\underline{d}_i$ is sampled from the same normal distribution. The problem is that $\underline{d}_i$ in our approach is not sampled from the same distribution unless we apply method C in both the pilot-phase and the final phase, and $M = N$.

The analysis of the simulation output also forms a complication in the following variants of the original problem formulation.

(i) Instead of minimizing the variance subject to the computer time restriction, we may want to solve the converse problem. So we minimize the computer time needed to simulate both systems given by

$$T = t_1 M + t_2 N \tag{302}$$

This minimization is restricted by the condition of a fixed reliability of the estimated difference. This reliability may be measured by the variance. Hence the minimization of (302) is subject to the condition

$$\mathrm{var}(\bar{\underline{d}}) = a_1 M^{-1} + a_2 N^{-1} = V \tag{303}$$

where V is a suitably chosen constant. If we know that we shall take so many observations that we can use (298) above, then we can take the constant

$$V = (\delta^* / z^{\alpha/2})^2 \tag{304}$$

supposing that we want to be $(1 - \alpha)\%$ certain that we do select the system with the highest mean if the best mean is at least δ^* units better than the mean of the inferior system. We refer to Chapter V for a further discussion of the determination of V. If we do not suppose beforehand that we shall take a great many observations then we have the same kind of problem as we discussed in relation with (299), and the determination of V in (303) becomes difficult. Once V is specified it can be found in Fishman (1967, p. 6) that the solution of (302) and (303) is given by (305) and

(306); this solution can be easily verified following the approach of Appendix 13.[63]

$$M' = \{a_1 + (t_1^{-1} t_2 a_1 a_2)^{1/2}\}/V \tag{305}$$

$$N' = \{a_2 + (t_1 t_2^{-1} a_1 a_2)^{1/2}\}/V \tag{306}$$

(ii) We may generalize our original problem to K (≥ 2) systems. Following Lombaers (1968, p. 254) we might be tempted to advise the following procedure. Suppose that among the K systems there is one "control" system, e.g., there are $(K - 1)$ new system variants and there is one existing system variant. Then simulate many runs (say M) of the control system and fewer runs (say N) of the $(K - 1)$ experimental systems; in Scheffé (1964, p. 88) it is stated that under certain assumptions the optimal ratio is $M/N = \sqrt{K - 1}$. The control system is generated applying antithetic variates; the $(K - 1)$ experimental systems are generated using the same random numbers as the control system. However, in this procedure the undesirable negative cross-covariance c_4 is again overlooked. For besides every positive cross-covariance c_3, a negative cross-covariance c_4 is created. So instead of this procedure we may use an optimization approach as presented above. Instead of minimizing $\text{var}(\bar{\underline{x}} - \bar{\underline{y}})$ we may now decide to minimize the sum of the variances

$$\sum_{k=2}^{K} \text{var}(\bar{\underline{x}}_1 - \bar{\underline{x}}_k) \tag{307}$$

where $\bar{\underline{x}}_1$ is the average of the runs of the control system and $\bar{\underline{x}}_k$ is the average for the experimental system k. If there is no control system, then we compare all the systems with each other, so we have $K(K - 1)/2$ comparisons. Then (307) may be replaced by

$$\sum_{i=1}^{K-1} \sum_{k=i+1}^{K} \text{var}(\bar{\underline{x}}_i - \bar{\underline{x}}_k) \tag{308}$$

Both (307) and (308) are constrained by the computer time restriction

$$\sum_{1}^{K} t_k N_k = T \tag{309}$$

Each of the variances after the summation-sign in (307) or (308) is given by expressions like the ones in Table 10. The restriction in (309) can be incorporated into the minimization of (307) and (308) by the method of Lagrangian multipliers. We would have to investigate more cases than in Table 10 since, besides comparing N_1 with N_2, we would have to compare N_1 with N_3, etc. So the optimization procedure can still be applied but it becomes more cumbersome. The statistical analysis also forms a serious problem. For, when analyzing the simulation experiment with K systems we may have to use "multiple comparison" procedures. As we shall see in Chapter V these procedures yield simultaneous confidence intervals for comparisons among the K systems. Unfortunately they are based on independent observations. Hence the methods B and C would conflict with this assumption. Method A does not conflict since we can take the averages of the antithetic pairs as the new observations as in (239). Next consider the converse problem. If we do not limit the available computer time and hence the number of observations, then we can continue the simulation until a desired reliability level is reached. For the case of two systems this was formulated in (302) and (303). If there are K systems then we may apply "multiple ranking" procedures.[64)] In Chapter V it will be shown that these procedures guarantee that the probability of selecting the best system is at least $(1 - \alpha)\%$. One of the assumptions of these procedures is again the independence of the observations. Method A is the only method that satisfies this assumption.

Above we saw that there are serious problems involved in the statistical analysis of the output of the two or more generally the K systems, either when minimizing the variance or when minimizing the computer time. Therefore we may decide not to use the above optimization procedure for the selection of one of the three methods A, B, C. Instead we may a priori choose method A, antithetic variates only. The above procedure is then still useful for the optimal allocation of the restricted computer time between the systems. From Appendix 12 it follows that the coefficients a_1 and a_2 are always positive for method A. Table 10 shows that if we use only method A we do not have to distinguish between $M \geq N$ and $M \leq N$. Hence the optimal values of M and N can be determined without using Table 11 and (277). Instead we use the integer-restrictions in (278) through (280) and Table 15.

TABLE 15. Optimal Values of M and N When Applying Antithetic Variates Only

Side conditions	M_0	N_0
$M^* \geq N_p$, $N^* \geq N_p$	M^*	N^*
$M^* < N_p$	N_p	$(T - t_1 N_p)/t_2$
$N^* < N_p$	$(T - t_2 N_p)/t_1$	N_p

III.9. SUMMARY AND CONCLUSIONS

In this chapter we saw that there are several VRT's that can be applied in the simulation of complex systems and that yield worthwhile variance reductions. However, two VRT's stand out because of their extreme simplicity, namely antithetic variates and common random numbers. They are simple to apply so little extra programming and running time is necessary, and yet they yield worthwhile variance reductions. Moreover they do not distort the sampling process and therefore besides the average response, the dynamic behavior of the system can be studied. (This dynamic behavior is interesting in itself and may also be studied for the validation of the model.) Of these two VRT's the use of the same random numbers is indeed a technique that is widely applied. Antithetic variates, notwithstanding their simplicity, have been hardly used in the simulation of complex systems.

It was shown that the joint application of common random numbers and antithetic variates does not necessarily result in the largest variance reduction. We derived a procedure for choosing that method from among (A) antithetic variates, (B) common random numbers, and (C) their joint application that yields the smallest variance. This procedure also gives the optimal allocation of the limited computer time for the comparison of two systems. Besides increasing the reliability of the simulation experiment we usually also want to measure this reliability. To measure the reliability we apply statistical techniques that are based on independent observations. Common random numbers violate this basic assumption; antithetic variates, however, still yield independent observations if we take the averages of the antithetic pairs.

Until now most simulation experiments have been performed without a formal statistical analysis of the results. We expect that in the future such an analysis will more often be performed in order to reach reliable conclusions. Hence we expect that less use will be made of common random numbers, since they conflict with the basic assumption of most statistical analysis procedures, and that more use will be made of antithetic variates.

III.10. LITERATURE

The handbook on VRT's to which many authors, e.g., Naylor (1971, p. 25) refer is Hammersley and Handscomb (1964). This book may be very useful for the application of VRT's in Monte Carlo problems, but for simulation, we consider it to be completely unsuited. We think that the dissertation by Moy (1965) gives a good survey of the use of VRT's in simulation. A summary of this book is Moy (1966).

APPENDIX III.1. THE JACKKNIFE STATISTIC

The purpose of jackknifing is to decrease the bias of an estimator and to obtain a confidence interval for the estimator. The procedure runs as follows. Suppose we have an estimator $\hat{\theta}$ based on n independent observations $x_1, \ldots, x_n$. Divide this sample into N groups of equal size $M = n/N$. Form a subsample by deleting one group from the N groups and calculate the same estimator from the remaining $(N - 1)M$ observations. Denote the estimator obtained by deleting group i by the symbol $\hat{\theta}_i$ $(i = 1, \ldots, N)$. Then the "pseudo values" J_i of the jackknife statistic are defined as

$$J_i = n\hat{\theta} - (N - 1)\,\hat{\theta}_i \qquad (i = 1, \ldots, N) \tag{1.1}$$

and the jackknife estimator of θ is

$$\bar{J} = \sum_{i=1}^{N} J_i/N = N\hat{\theta} - (N - 1)\,\bar{\hat{\theta}}_i \tag{1.2}$$

This jackknife statistic was introduced by Quenouille (1956) to reduce possible bias of the estimator $\hat{\theta}$. For, if θ has bias of order n^{-1} then in many cases $\bar{J}$ has bias of order n^{-2}; see Gray and Schucany (1972, p. 7). Later Tukey (1958) suggested that the pseudovalues J_i could be treated as independent identically distribued variables and hence approximate confidence intervals can be derived for $\bar{J}$ from the t-distribution with $N - 1$ degrees of freedom:

$$t_{N-1} \approx \frac{(\bar{J} - \theta)\, N^{1/2}}{\{\Sigma_i(J_i - \bar{J})^2/(N - 1)\}^{1/2}} \tag{1.3}$$

We refer to Gray and Schucany (1972, pp. 139, 149, 158) for proofs of the asymptotic normality of the right-hand side in (1.3) under various conditions. Note that if in a particular problem the J_i can assume only $N_1 < N$ distinct values then the degrees of freedom are put equal to $N_1 - 1$; see Gray and Schucany (1972, p. 164). If the $\underline{J}_i$ are not independent but instead all pairs have positive correlation ρ (e.g., $\rho = 1/n$) then the right-hand side of (1.3) should be multiplied by the correction factor

$$f = \left[\frac{1 - \rho}{1 + (N - 1)\rho} \right]^{1/2} \tag{1.4}$$

as stated by Gray and Schucany (1972, pp. 165, 173-174). As a conservative heuristic rule-requiring more research we would propose: Estimate ρ (> 0) and substitute $\hat{\underline{\rho}}$ into (1.4). It is further desirable to reduce possible skewness of the $\hat{\underline{\theta}}_i$ by an appropriate transformation, e.g., a logarithmic one for $\hat{\underline{\theta}} = \hat{\underline{\sigma}}^2$; see Schucany and Gray (1972, p. 169).

Many applications of the jackknife can be found in Gray and Schucany; see also Arvesen and Salsburg (1975) and Goodman et al. (1973). The jackknife statistic (1.2) can be generalized to

$$\underline{G} = \frac{\hat{\underline{\theta}}_1 - R\hat{\underline{\theta}}_2}{1 - R} \qquad (R \neq 1) \tag{1.5}$$

where $\hat{\underline{\theta}}_1$ and $\hat{\underline{\theta}}_2$ are estimators of θ. Obviously (1.5) reduces to (1.2) if we put

$$R = \frac{N - 1}{N}, \qquad \hat{\underline{\theta}}_1 = \hat{\underline{\theta}}, \qquad \hat{\underline{\theta}}_2 = \hat{\underline{\theta}}_i \tag{1.6}$$

We refer to Arvesen and **Salsburg** (1972) and Gray and Schucany (1972) for a detailed exposé on the (original and generalized) jackknife, and for additional references. The jackknife may also be used when $\hat{\underline{\theta}}$ is based not only on the $\underline{x}$ themselves but also on concomitant variables $\underline{y}_1 \ldots \underline{y}_n$; see Quenouille (1956, pp. 357-358) and our section on control variates.

APPENDIX III.2. STRATIFICATION AFTER SAMPLING WITH COMBINATION OF STRATA

In this appendix we shall show that bias is created when we use an estimator based on stratification after sampling with combination of strata if a stratum remains empty. This estimator $\bar{\underline{x}}_{SAC}$ was defind in (22) through (24) in Section III.2.2. Consider the expected value of $\bar{\underline{w}}_k$ defined in (23a) and (23b) above.

$$E(\bar{\underline{w}}_k) = \sum_{i=0}^{n} E(\bar{\underline{w}}_k | \underline{n}_k = i)\, P(\underline{n}_k = i)$$

$$= \sum_{j=1}^{n} E(\bar{\underline{w}}_k | \underline{n}_k = j)\, P(\underline{n}_k = j) + E(\bar{\underline{w}}_k | \underline{n}_k = 0)\, P(\underline{n}_k = 0) \qquad (2.1)$$

From (23a) it follows that

$$E(\bar{\underline{w}}_k | \underline{n}_k = j) = E(\sum_{h=1}^{j} \underline{x}_{kh}/j)$$

$$= \sum_{h=1}^{j} E(\underline{x}_{kh})/j$$

$$= \sum_{h=1}^{j} \mu_k/j = \mu_k \qquad (2.2)$$

and from (23b) it follows that

$$E(\bar{\underline{w}}_k | \underline{n}_k = 0) = E\left(\sum_{h=1}^{\underline{n}_g} \underline{x}_{gh}/\underline{n}_g\right) \qquad (2.3)$$

where

$$\underline{n}_g > 0 \qquad (2.4)$$

Hence

$$E(\underline{\bar{w}}_k | \underline{n}_k = 0) = \sum_{j=1}^{n} E\left(\sum_{h=1}^{\underline{n}_g} \underline{x}_{gh}/\underline{n}_g | \underline{n}_g = j \right) P(\underline{n}_g = j)$$

$$= \sum_{j=1}^{n} E(\sum_{h=1}^{j} \underline{x}_{gh}/j)\ P(\underline{n}_g = j)$$

$$= \sum_{j=1}^{n} \mu_g\ P(\underline{n}_g = j)$$

$$= \mu_g \sum_{j=1}^{n} P(\underline{n}_g = j) = \mu_g \qquad (2.5)$$

where the last equality holds since

$$\sum_{j=1}^{n} P(\underline{n}_g = j) = 1 \qquad (2.6)$$

because of (2.4). Substitution of (2.2) and (2.5) into (2.1) yields

$$E(\underline{\bar{w}}_k) = \sum_{j=1}^{n} \mu_k\ P(\underline{n}_k = j) + \mu_g\ P(\underline{n}_k = 0)$$

$$= \mu_k \sum_{j=1}^{n} P(\underline{n}_k = j) + \mu_g\ P(\underline{n}_k = 0) \qquad (2.7)$$

If

$$\mu_g = \mu_k \qquad (2.8)$$

or

$$P(\underline{n}_k = 0) = 0 \qquad (2.9)$$

held then we could replace (2.7) by

$$E(\underline{\bar{w}}_k) = \mu_k \sum_{j=0}^{n} P(\underline{n}_k = j) = \mu_k 1 = \mu_k \qquad (2.10)$$

which, because of (22), would yield

$$E(\underline{\bar{x}}_{SAC}) = \sum_{k=1}^{K} p_k \mu_k = \mu \qquad (2.11)$$

So if neither (2.8) nor (2.9) holds then the stratified estimator $\underline{\bar{x}}_{SAC}$ is biased.

APPENDIX III.3. THE MEAN SQUARE ERROR

Suppose we want to estimate a population parameter μ by the statistic $\underline{y}$. So μ may be the population mean, $\underline{y}$ the sample average or median, etc. Denote the expected value of $\underline{y}$ by η, i.e.,

$$E(\underline{y}) = \eta \tag{3.1}$$

Then the bias of the estimator $\underline{y}$ is defined by

$$\text{BIAS} = E(\underline{y}) - \mu = \eta - \mu \tag{3.2}$$

The variance of $\underline{y}$ is defined by

$$\sigma^2 = \text{var}(\underline{y}) = E[\{\underline{y} - E(\underline{y})\}^2] = E[(\underline{y} - \eta)^2] \tag{3.3}$$

The mean square error, or briefly the MSE, is the expected squared deviation between $\underline{y}$ and the parameter it is supposed to estimate. So

$$\begin{aligned} \text{MSE}(\underline{y}) &= E[(\underline{y} - \mu)^2] \\ &= E[\{(\underline{y} - \eta) + (\eta - \mu)\}^2] \\ &= E[(\underline{y} - \eta)^2 + (\eta - \mu)^2 + 2(\underline{y} - \eta)(\eta - \mu)] \\ &= \text{var}(\underline{y}) + (\text{BIAS})^2 \end{aligned} \tag{3.4}$$

Note that we can estimate the MSE by taking, say, N samples (of the same size), determining for each sample the statistic $\underline{y}_i$ $(i = 1, \ldots, N)$ and calculating the estimator given in (3.5) assuming that we know μ.

$$\begin{aligned} \underline{\widehat{\text{MSE}}} &= \sum_{i=1}^{N} (\underline{y}_i - \mu)^2/N \\ &= \sum_{i=1}^{N} \{(\underline{y}_i - \bar{\underline{y}}) + (\bar{\underline{y}} - \mu)\}^2/N \\ &= \sum_{i=1}^{N} (\underline{y}_i - \bar{\underline{y}})^2/N + \sum_{i=1}^{N} (\bar{\underline{y}} - \mu)^2/N + \sum_{i=1}^{N} 2(\underline{y}_i - \bar{\underline{y}})(\bar{\underline{y}} - \mu)/N \\ &= \underline{\hat{\sigma}}^2 + (\bar{\underline{y}} - \mu)^2 \end{aligned} \tag{3.5}$$

where $\bar{\underline{y}}$ is the average value of the estimator in the N replicated samples, i.e.,

$$\bar{\underline{y}} = \sum_{i=1}^{N} \underline{y}_i / N \tag{3.6}$$

and $\hat{\underline{\sigma}}^2$ is the estimator of the variance of $\underline{y}$ given in (3.7).

$$\hat{\underline{\sigma}}^2 = \sum_{i=1}^{N} (\underline{y}_1 - \bar{\underline{y}})^2 / N \tag{3.7}$$

The expected value of $\hat{\underline{\sigma}}^2$ can be found in e.g., Fisz (1967, p. 347) and is

$$E(\hat{\underline{\sigma}}^2) = \frac{N - 1}{N} \sigma^2 \tag{3.8}$$

The expected value of the squared estimated bias can be easily proved to be

$$E(\bar{\underline{y}} - \mu)^2 = \frac{\sigma^2}{N} + (\text{BIAS})^2 \tag{3.9}$$

Combining (3.8) and (3.9) shows that (3.5) is an unbiased estimator of the MSE. For

$$\begin{aligned} E(\widehat{M\underline{S}E}) &= E(\hat{\underline{\sigma}}^2) + E[(\bar{\underline{y}} - \mu)^2] \\ &= \frac{N - 1}{N} \sigma^2 + \frac{1}{N} \sigma^2 + (\text{BIAS})^2 \\ &= \sigma^2 + (\text{BIAS})^2 \end{aligned} \tag{3.10}$$

APPENDIX III.4. THE ESTIMATED VARIANCE IN STRATIFICATION AFTER SAMPLING

In stratification after sampling the number of observations per stratum is a stochastic variable. In order to derive the variance of the stratified estimator, we use (4.1); this relation is proved in Keeping (1962, p. 398-399).

$$\operatorname{var}(\underline{x}) = \underset{\underline{y}}{E}\,[\operatorname{var}(\underline{x}|y)] + \underset{\underline{y}}{\operatorname{var}}\,[E(\underline{x}|y)] \tag{4.1}$$

Let us consider the variance of $\bar{\underline{v}}_k$, the stratum average if we continue sampling until each stratum is nonempty, as specified in (29) through (32) in Section III.2.2.

$$\operatorname{var}(\bar{\underline{v}}_k) = \underset{\underline{n}_k}{E}\left[\operatorname{var}\left(\sum_{j=1}^{\underline{n}_k} \underline{x}_{kj}/\underline{n}_k \,|\, \underline{n}_k = n_k\right)\right] + \underset{\underline{n}_k}{\operatorname{var}}\left[E\left(\sum_{j=1}^{\underline{n}_k} \underline{x}_{kj}/\underline{n}_k \,|\, \underline{n}_k = n_k\right)\right]$$

$$= \underset{\underline{n}_k}{E}\,[\operatorname{var}(\underline{x}_{kj})/n_k] + \underset{\underline{n}_k}{\operatorname{var}}\,[\mu_k]$$

$$= \operatorname{var}(\underline{x}_{kj})\, E(1/\underline{n}_k) + 0 \tag{4.2}$$

where we point out that both $\operatorname{var}(\underline{x}_{kj})$ and μ_k are nonstochastic. Consider the estimator of $\operatorname{var}(\bar{\underline{v}}_k)$ in (4.3).

$$\widehat{\operatorname{var}}(\bar{\underline{v}}_k) = \frac{\sum_{j=1}^{\underline{n}_k} (\underline{x}_{kj} - \bar{\underline{v}}_k)^2}{(\underline{n}_k - 1)} \frac{1}{\underline{n}_k} \tag{4.3}$$

Observe that (4.3) implies that we replace the original condition (31) by the modified condition

$$\underline{n}_k > 1 \quad \text{for all } k \tag{4.4}$$

Then (4.3) is an unbiased estimator of (4.2) since

$$E[\hat{\underline{var}}(\bar{\underline{v}}_k)] = \sum_{n_k} E[\hat{\underline{var}}(\bar{\underline{v}}_k)|\underline{n}_k = n_k)]\ P(\underline{n}_k = n_k)$$

$$= \sum_{n_k} E\left(\frac{\Sigma_{j=1}^{n_k}(\underline{x}_{kj} - \bar{\underline{v}}_k)^2}{(n_k - 1)}\ \frac{1}{n_k}\right) P(\underline{n}_k = n_k)$$

$$= \sum_{n_k} \frac{var(\underline{x}_{kj})}{n_k}\ P(\underline{n}_k = n_k)$$

$$= var(\underline{x}_{kj})\ E(1/\underline{n}_k) \qquad (4.5)$$

From (29) it follows that the stratified estimator $\bar{\underline{x}}_{SAP}$ is a weighted average of the independent stratum-averages $\bar{\underline{v}}_k$. Hence its variance is given by

$$var(\bar{\underline{x}}_{SAP}) = \sum_{k=1}^{K} p_k^2\ var(\bar{\underline{v}}_k) \qquad (4.6)$$

Substituting (4.3), the unbiased estimator of $var(\bar{\underline{v}}_k)$, into (4.6) yields an unbiased estimator of $var(\bar{\underline{x}}_{SAP})$.

APPENDIX III.5. AN EXAMPLE OF THE GENERATION OF RANDOM NUMBERS IN STRATIFIED SAMPLING

In this appendix we shall demonstrate the rationale of Moy's procedure for the generation of a vector of m random numbers when g random numbers should be not smaller than the constant c. Moy's procedure was given in Fig. 2 in Section III.2.4. Suppose that $m = 4$ and $g = 2$. Then we have $\binom{4}{2} = 6$ different vectors as shown in Table 4.1. From this table we see

$$P(\underline{r}_1 \geq c) = 3/6 = g/m = 2/4 \qquad \text{(compare vectors 1, 2, and 3)}$$

$$P(\underline{r}_2 \geq c|\underline{r}_1 \geq c) = 1/3 = (g - 1)/(m - 1) \quad \text{(compare vectors 1, 2, and 3)}$$

$$P(\underline{r}_2 \geq c|\underline{r}_1 < c) = 2/3 = g/(m - 1) \qquad \text{(compare vectors 4, 5, and 6)}$$

$$P(\underline{r}_3 \geq c|\underline{r}_1 \geq c \text{ and } \underline{r}_2 \geq c) = 0 = (g - 2)/(m - 2) \qquad \text{(compare vector 1)}$$

$$P(\underline{r}_3 \geq c|\underline{r}_1 \geq c \text{ and } \underline{r}_2 < c) = 1/2 = (g - 1)/(m - 2) \qquad \text{(compare vectors 2 and 3)}$$

TABLE 4.1. The Six Different Vectors When $m = 4$ and $g = 2$[a]

Vector	r_1	r_2	r_3	r_4
1	*	*		
2	*		*	
3	*			*
4		*	*	
5		*		*
6			*	*

[a] An asterisk denotes $r \geq c$.

APPENDIX III.6. EXPECTED VALUE OF A CONTROL VARIATE WITH STOCHASTIC $\underline{m}$

The control variate is

$$b_1 + b_2 \sum_1^{\underline{m}} \underline{y}_t + b_3 \left(\sum_1^{\underline{m}} \underline{y}_t \right)^2 \tag{6.1}$$

Using the symbols $\eta_1 = E(\underline{y}_t)$, $\eta_2 = E(\underline{y}_t^2)$ and $\sigma_1^2 = \eta_2 - \eta_1^2 = \operatorname{var}(\underline{y}_t)$ we obtain

$$E\left(\sum_1^{\underline{m}} \underline{y}_t \right) = \underset{\underline{m}}{E} \left[E\left(\sum_1^{\underline{m}} \underline{y}_t \middle| \underline{m} = m \right) \right] = \underset{\underline{m}}{E}[m\eta_1] = \eta_1 E(\underline{m}) \tag{6.2}$$

and

$$E\left[\left(\sum_1^{\underline{m}} \underline{y}_t \right)^2 \right] = \underset{\underline{m}}{E} \left[E\left\{ \left(\sum_1^{\underline{m}} \underline{y}_t \right)^2 \middle| \underline{m} = m \right\} \right] \tag{6.3}$$

where

$$\begin{aligned} E\left[\left(\sum_1^{m} \underline{y}_t \right)^2 \right] &= E\left[\left(\sum_{t=1}^{m} \underline{y}_t \right) \left(\sum_{s=1}^{m} \underline{y}_s \right) \right] \\ &= E\left[\sum_{t=1}^{m} \sum_{s=1}^{m} \underline{y}_t \, \underline{y}_s \right] \\ &= \sum_{t \neq s} \sum E(\underline{y}_t \, \underline{y}_s) + \sum_{t=1}^{m} E(\underline{y}_t^2) \end{aligned}$$

$$= \sum_{t \neq s} \sum E(\underline{y}_t)\ E(\underline{y}_s) + m\eta_2$$

$$= m(m - 1)\ \eta_1^2 + m\eta_2$$

$$= m^2\eta_1^2 + m(\eta_2 - \eta_1^2)$$

$$= m^2\eta_1^2 + m\sigma_1^2 \qquad (6.4)$$

In (6.4) η_1 and σ_1 can be calculated from the distribution of the input variable $\underline{y}_t$. Substitution of (6.4) into (6.3) yields

$$E\left[\left(\sum_1^{\underline{m}} \underline{y}_t\right)^2\right] = \underset{\underline{m}}{E}[\eta_1^2\ \underline{m}^2 + \sigma_1^2\ \underline{m}]$$

$$= \eta_1^2\ E(\underline{m}^2) + \sigma_1^2\ E(\underline{m}) \qquad (6.5)$$

APPENDIX III.7. THE COVARIANCES AMONG CONTROL VARIATE ESTIMATORS

Define the control variate estimators as in (141), i.e.,

$$\underline{x}cze_{ji} = \underline{x}_{ji} - \sum_{k=1}^{K} \hat{\underline{a}}_{kj}\ \underline{z}_{kji} \qquad (7.1)$$

Consider the covariance between two estimators <u>within</u> group j, i.e.,

$$\text{cov}(\underline{x}cze_{ji}, \underline{x}cze_{jh})$$

$$= E\left[\{\underline{x}_{jk} - \sum_k \hat{\underline{a}}_{kj}\ \underline{z}_{kji} - \mu\}\ \{\underline{x}_{jh} - \sum_k \hat{\underline{a}}_{kj}\ \underline{z}_{kjh} - \mu\}\right] \qquad (7.2)$$

where μ is the expectation of $\underline{x}cze_{ji}$, this expectation being equal to that of $\underline{x}_{ji}$ since both are unbiased estimators. So

$$\text{cov}(\underline{x}cze_{ji}, \underline{x}cze_{jh})$$

$$= E\left[(\underline{x}_{ji} - \mu)(\underline{x}_{jh} - \mu)\right] - E\left[(\underline{x}_{ji} - \mu)\ (\sum_k \hat{\underline{a}}_{kj}\ \underline{z}_{kjh})\right]$$

$$- E\left[(\sum_k \hat{\underline{a}}_{kj}\ \underline{z}_{kji})(\underline{x}_{jh} - \mu)\right] + E\left[(\sum_k \hat{\underline{a}}_{kj}\ \underline{z}_{kji})(\sum_k \hat{\underline{a}}_{kj}\ \underline{z}_{kjh})\right] \qquad (7.3)$$

The first term in (7.3) is the covariance between two independent r replications of a simulation run; hence this term is zero. In the second term the first factor, $(\underline{x}_{ji} - \mu)$, is independent of the second factor since $\hat{\underline{a}}_{kj}$ is estimated from observations outside group j, and $\underline{z}_{kjh}$ refers to run h which is independent of run i, both i and h being within group j. Hence

$$E\left[(\underline{x}_{ji} - \mu)(\sum_k \hat{\underline{a}}_{kj} \underline{z}_{kjh})\right] = E\left[(\underline{x}_{ji} - \mu)\right] E\left[(\sum_k \hat{\underline{a}}_{kj} \underline{z}_{kjh})\right] = 0 \tag{7.4}$$

In the same way we derive that the third term in (7.3) is zero. The fourth term can be written as

$$E[\sum_k \sum_{k'} \hat{\underline{a}}_{kj} \underline{z}_{kji} \hat{\underline{a}}_{k'j} \underline{z}_{k'jh}]$$

$$= \sum_k \sum_{k'} E(\hat{\underline{a}}_{kj} \hat{\underline{a}}_{k'j}) E(\underline{z}_{kji}) E(\underline{z}_{k'jh}) \tag{7.5}$$

since $\hat{\underline{a}}_{kj}$ and $\hat{\underline{a}}_{k'j}$ are estimated from runs outside group j, while further $\underline{z}_{kji}$ and $\underline{z}_{k'jh}$ are independent since runs i and h are independent. As the control variates have zero-expectation the fourth term in (7.3) is zero. Summarizing, the covariance between two estimators <u>within</u> the same group is zero.

Next consider the covariance between two observations in different groups, say observation i in group j and observation h in group p. Analogous to (7.3) we derive

$$\text{cov}(\underline{x}cze_{ji}, \underline{x}cze_{ph})$$

$$= E[(\underline{x}_{ji} - \mu)(\underline{x}_{ph} - \mu)] - E\left[(\underline{x}_{ji} - \mu)(\sum_k \hat{\underline{a}}_{kp} \underline{z}_{kph})\right]$$

$$- E\left[(\sum_k \hat{\underline{a}}_{kj} \underline{z}_{kji})(\underline{x}_{ph} - \mu)\right] + E\left[(\sum_k \hat{\underline{a}}_{kj} \underline{z}_{kji})(\sum_k \hat{\underline{a}}_{kp} \underline{z}_{kph})\right] \tag{7.6}$$

The first term is zero again. The second term becomes

$$E\left[\sum_k \hat{\underline{a}}_{kp} (\underline{x}_{ji} - \mu) \underline{z}_{kph}\right] = \sum_k E[\hat{\underline{a}}_{kp} (\underline{x}_{ji} - \mu)] E(\underline{z}_{kph}) = 0 \tag{7.7}$$

In the same way we see that the third term is also zero. The last term is equal to

$$\sum_k \sum_{k'} E(\hat{\underline{a}}_{kj} \underline{z}_{kji} \hat{\underline{a}}_{kp} \underline{z}_{kph}) \tag{7.8}$$

where $\underline{z}_{kji}$ cannot be taken apart since it is not independent of $\hat{\underline{a}}_{kp}$. Hence the covariance between two estimators belonging to two different groups, is not zero.

APPENDIX III.8. GAMMA APPROXIMATION FOR THE OPTIMAL DENSITY FUNCTION IN THE IMPORTANCE SAMPLING EXAMPLE

Figure 3 in Section III.5.2 suggests that we may approximate $h_0(x)$ by a continuous distribution with a modus smaller than its mean. (The modus is the value of the stochastic variable that has the highest probability.) Most standard distributions, however, are either discrete or symmetric. Yet the gamma distribution has the desired characteristics. Its density function is

$$G(x) = \frac{b^p}{\Gamma(p)} x^{p-1} e^{-bx} \qquad (x, b, p > 0) \tag{8.1}$$

After having selected the type of distribution, we have to fix the values of the parameters. Generating a value of $\underline{x}$ from a gamma distribution is simple if the parameter p in (8.1) is an integer. For Naylor et al. (1967a, p. 88) show that such a gamma distributed variable can be generated by taking the sum of p exponentially distributed variables. Further it is known, see Keeping (1962, p. 81), that the gamma distribution becomes more asymmetric as p decreases. The smallest integer $p = 1$, however, yields the exponential distribution that decreases continuously for increasing x. Comparison with Fig. 3 in Section III.5.2 shows that we want a density function that does not start decreasing immediately as x increases; instead the function should have a "top" for $x > 0$.

Therefore we choose the next smallest integer, i.e., $p = 2$. So the new density function, $h_2(x)$, follows from substitution of $p = 2$ into (8.1) and is given by

$$h_2(x) = b^2 \, xe^{-bx} \qquad (x, b > 0) \tag{8.2}$$

From (158), (164), (165) and (8.2) follows

$$\begin{aligned} g_2^*(x) &= 0 && \text{if} \quad 0 < x < v \\ &= \frac{x^{-1} \lambda e^{-\lambda x}}{b^2 \, xe^{-bx}} = \frac{\lambda}{b^2} x^{-2} \, e^{-(\lambda-b)x} && \text{if} \quad x \geq v \end{aligned} \tag{8.3}$$

In Kleijnen (1968, pp. 197-198) it is proved that we cannot calculate the value of b that minimizes the variance of $g_2^*(\underline{x})$ since we would need to know ξ which is to be estimated. Therefore, as we explained in Section III.5.2 we minimize the range of $g_2^*(\underline{x})$. Consequently to keep this range finite we take $b \leq \lambda$. Hence

$$\begin{aligned} R[g_2^*(\underline{x})] &= \max_x [g_2^*(x)] - \min_x [g_2^*(x)] \\ &= g_2^*(v) - 0 \\ &= \frac{\lambda}{b^2} v^{-2} \, e^{-(\lambda-b)v} \end{aligned} \tag{8.4}$$

Next the value of b that minimizes R is found differentiating R with respect to b, putting the result equal to zero and solving for b. This procedure yields $b = 2/v$. This value of b can be substituted into (8.2) and (8.3) to find the new density function $h_2(x)$ and the estimator $g_2^*(\underline{x})$ in the gamma approximation. Note that the new density function has modus $v/2$ and expectation v. Further we point out that (8.4) is based on the condition $b \leq \lambda$ which may conflict with $b = 2/v$. We refer to Kleijnen (1968b, pp. 185-186) for a discussion of this point. There it is shown that there are no serious problems involved. As in (174) and (175) we can prove that the range of $g_2^*(\underline{x})$ is smaller than that of $g(\underline{x})$, the crude estimator. The variance of $g_2^*(\underline{x})$ is discussed in Section III.5.2.

APPENDIX III.9. THE VARIANCE OF THE IMPORTANCE SAMPLING ESTIMATOR $g_3^*(\underline{x})$

The importance sampling estimator $g_3^*(\underline{x})$ is given by

$$g_3^*(\underline{x}) = \frac{1}{x} \lambda v \exp[-(\lambda - \frac{1}{v})x] \qquad \text{if} \quad x \geq v$$

$$= 0 \qquad \text{if} \quad 0 \leq x < v \tag{9.1}$$

The variance of $g_3^*(\underline{x})$ is calculated from

$$\text{var}[g_3^*(\underline{x})] = E[\{g_3^*(\underline{x})\}^2] - E[g_3^*(\underline{x})]\, E[g_3^*(\underline{x})] \tag{9.2}$$

Since $g_3^*(\underline{x})$ is an unbiased estimator of $\xi(\lambda,v)$ we have

$$E[g_3^*(\underline{x})] = \xi(\lambda,v) \tag{9.3}$$

Further

$$E[\{g_3^*(\underline{x})\}^2] = \int_{-\infty}^{\infty} \{g_3^*(x)\}^2\, h_3(x)\, dx \tag{9.4}$$

where $h_3(x)$ is the new density function of $\underline{x}$, i.e.,

$$h_3(x) = \frac{1}{v} \exp(-\frac{1}{v} x) \qquad \text{if} \quad x \geq 0$$

$$= 0 \qquad \text{if} \quad x < 0 \tag{9.5}$$

Hence

$$E[\{g_3^*(\underline{x})\}^2] = \int_v^{\infty} \frac{1}{x^2} (\lambda v)^2 \exp[-2(\lambda - \frac{1}{v})x]\, \frac{1}{v} \exp(-\frac{1}{v} x)\, dx \tag{9.6}$$

$$= \lambda^2 v \int_v^{\infty} \frac{1}{x^2} \exp[-(2\lambda - \frac{1}{v})x]\, dx \tag{9.6}$$

Partial integration gives

$$E[\{g_3^*(\underline{x})\}^2] = \lambda^2[e^{(1-2\lambda v)} + (1 - 2\lambda v) \int_v^{\infty} \frac{1}{x} \exp[-(2\lambda - \frac{1}{v})x]\, dx \tag{9.7}$$

provided that $\lambda > 2/v$ since otherwise the integral diverges. Using the symbol Ψ where

$$\Psi = -2\lambda + \frac{1}{v} \tag{9.8}$$

and the definition of $\xi(\lambda,v)$ in (163) yields

$$E[\{g_3^*(\underline{x})\}^2] = \lambda^2[e^{\Psi v} - v\ \xi(-\Psi,v)] \tag{9.9}$$

Hence

$$\text{var}[g_3^*(\underline{x})] = \lambda^2[e^{\Psi v} - v\ \xi(-\Psi,v)] - \{\xi(\lambda,v)\}^2 \tag{9.10}$$

APPENDIX III.10. GENERATING ANTITHETIC RANDOM NUMBERS FROM A MULTIPLICATIVE CONGRUENTIAL RELATION

The multiplicative congruential relation for generating random numbers is given by

$$x_i = ax_{i-1}\ (\text{mod } m) \qquad (i = 1, 2, \ldots) \tag{10.1}$$

where a is an integer and x_0 is a constant smaller than m. The <u>antithetic</u> random numbers r_i follow from

$$r_i = 1 - \frac{x_i}{m} \tag{10.2}$$

We can also generate identically the same antithetic random numbers from (10.3) through (10.5).

$$y_i = ay_{i-1}\ (\text{mod } m) \tag{10.3}$$

$$y_0 = m - x_0 \tag{10.4}$$

$$r_i' = \frac{y_i}{m} \tag{10.5}$$

In order to show that r_i and r_i' are identical, we proceed as follows.[65] We shall first prove that

$$x_i + y_i = m \qquad \text{for every } i \tag{10.6}$$

From (10.1) it follows that

$$x_i = ax_{i-1} \pmod{m} = ax_{i-1} - pm \tag{10.7}$$

where p is an integer, zero not being excluded. In the same way we have from (10.3)

$$y_i = ay_{i-1} \pmod{m} = ay_{i-1} - qm \tag{10.8}$$

Adding (10.7) and (10.8) yields

$$x_i + y_i = a(x_{i-1} + y_{i-1}) - (p + q)m \tag{10.9}$$

If

$$x_{i-1} + y_{i-1} = m \tag{10.10}$$

then (10.9) transforms to

$$x_i + y_i = (a - p - q)m \tag{10.11}$$

which means that $x_i + y_i$ is a multiple of m. Since, however, $0 < x_i < m$ and $0 < y_i < m$, (10.12) must hold.

$$x_i + y_i = m \tag{10.12}$$

So we have:

$$\text{If} \quad x_{i-1} + y_{i-1} = m \quad \text{then} \quad x_i + y_i = m \tag{10.13}$$

From (10.4) it follows that

$$x_0 + y_0 = m \tag{10.14}$$

Hence applying (10.13) for $i = 1$ we have

$$x_1 + y_1 = m \tag{10.15}$$

Taking $i = 2$ gives

$$x_2 + y_2 = m \tag{10.16}$$

etc. So in general

$$x_i + y_i = m \quad (i = 1, 2, \ldots) \tag{10.17}$$

Next we show that

$$r_i = r_i' \tag{10.18}$$

or

$$r_i - r_i' = 0 \tag{10.19}$$

Applying (10.2) and (10.5) we have

$$r_i - r_i' = 1 - \frac{x_i}{m} - \frac{y_i}{m} \tag{10.20}$$

Substitution of (10.7) and (10.8) into (10.20) yields

$$r_i - r_i' = 1 - \frac{ax_{i-1} - pm}{m} - \frac{ay_{i-1} - qm}{m}$$

$$= 1 + p + q - \frac{a(x_{i-1} + y_{i-1})}{m} \tag{10.21}$$

Using (10.17) results in

$$r_i - r_i' = 1 + p + q - a \tag{10.22}$$

Since p, q and a are integers while $0 < r_i < 1$ and $0 < r_i' < 1$ we have

$$r_i - r_i' = 0 \tag{10.23}$$

APPENDIX III.11. THE VARIANCE OF THE ESTIMATED DIFFERENCE BETWEEN THE RESPONSES OF TWO SYSTEMS

The difference between the responses of the two systems is estimated by

$$\underline{\bar{d}} = \sum_1^M \underline{x}_i/M - \sum_1^N \underline{y}_j/N \tag{11.1}$$

So

$$\text{var}(\underline{\bar{d}}) = \text{var}(\Sigma\, \underline{x}_i/M) + \text{var}(\Sigma\, \underline{y}_j/N) - 2\ \text{cov}(\Sigma\, \underline{x}_i/M,\ \Sigma\, \underline{y}_j/N)$$

$$= M^{-2}\ \text{var}(\Sigma\, \underline{x}_i) + N^{-2}\ \text{var}(\Sigma\, \underline{y}_j) - 2M^{-1}N^{-1}\ \text{cov}(\Sigma\, \underline{x}_i,\ \Sigma\, \underline{y}_j) \tag{11.2}$$

Remember that

$$\text{var}(\sum_i \underline{x}_i) = E\left[\{\sum_i \underline{x}_i - E(\sum_i \underline{x}_i)\}^2\right]$$

$$= E\left[\{\sum_i (\underline{x}_i - E(\underline{x}_i))\}^2\right]$$

$$= E\left[\sum_i \sum_g \{\underline{x}_i - E(\underline{x}_i)\}\{\underline{x}_g - E(\underline{x}_g)\}\right]$$

$$= \sum_i \sum_g E[\{\underline{x}_i - E(\underline{x}_i)\}\{\underline{x}_g - E(\underline{x}_g)\}]$$

$$= \sum_i \sum_g \text{cov}(\underline{x}_i,\ \underline{x}_g) \tag{11.3}$$

Hence

$$\text{var}(\underline{\bar{d}}) = M^{-2} \sum_{i=1}^{M} \sum_{g=1}^{M} \text{cov}(\underline{x}_i,\ \underline{x}_g) + N^{-2} \sum_{j=1}^{N} \sum_{h=1}^{N} \text{cov}(\underline{y}_j,\ \underline{y}_h)$$

$$- 2M^{-1}N^{-1} \sum_{i=1}^{M} \sum_{j=1}^{N} \text{cov}(\underline{x}_i,\ \underline{y}_j)$$

$$= M^{-2} \sum_{i}^{M} \sum_{\neq g}^{M} \text{cov}(\underline{x}_i,\ \underline{x}_g) + M^{-2} \sum_{i}^{M} \text{var}(\underline{x}_i)$$

$$+ N^{-2} \sum_{j}^{N} \sum_{\neq h}^{N} \text{cov}(\underline{y}_j,\ \underline{y}_h) + N^{-2} \sum_{j}^{N} \text{var}(\underline{y}_j)$$

$$+ 2M^{-1}N^{-1} \sum_{i}^{M} \sum_{j}^{N} \text{cov}(\underline{x}_i,\ \underline{y}_j) \tag{11.4}$$

where we may replace $\text{var}(\underline{x}_i)$ by σ_1^2 since all $\underline{x}_i$ have the same variance. Likewise $\text{var}(\underline{y}_j) = \sigma_2^2$. This gives

$$\mathrm{var}(\underline{\bar{d}}) = M^{-2} \sum_{i}^{M} \underset{\neq}{} \sum_{g}^{M} \mathrm{cov}(\underline{x}_i, \underline{x}_g) + M^{-1} \sigma_1^2$$

$$+ N^{-2} \sum_{j}^{N} \underset{\neq}{} \sum_{h}^{N} \mathrm{cov}(\underline{y}_j, \underline{y}_h) + N^{-1} \sigma_2^2$$

$$- 2M^{-1}N^{-1} \sum_{i}^{M} \sum_{j}^{N} \mathrm{cov}(\underline{x}_i, \underline{y}_j) \qquad (11.5)$$

APPENDIX III.12. THE SIGNS OF THE COEFFICIENTS a_1 AND a_2

In this appendix we shall study the signs of the coefficients a_1 and a_2 of $\mathrm{var}(\underline{\bar{d}})$ in (274). From Table 10 it follows that we have to investigate three cases.

(i) $\underline{\sigma_1^2 + c_1}$. From the definition of σ_1^2 and c_1 in (250) and (255), respectively, it follows that $(\sigma_1^2 + c_1)$ is negative if

$$\mathrm{var}(\underline{x}_i) < - \mathrm{cov}(\underline{x}_{2i}, \underline{x}_{2i-1}) \qquad (i = 1, 2, \ldots) \qquad (12.1)$$

or

$$\rho_1 < -1 \qquad (12.2)$$

where ρ_1 denotes the negative correlation coefficients of $\underline{x}_{2i}$ and $\underline{x}_{2i-1}$. Since a correlation coefficient must have a value in the interval $[-1,1]$ we conclude that $(\sigma_1^2 + c_1)$ is always positive. In the same way it can be proved that $(\sigma_2^2 + c_2)$ is always positive.

(ii) $\underline{\sigma_1^2 - 2c_3}$. Using the definition of c_3 in (257) we see that $(\sigma_1^2 - 2c_3)$ is negative if

$$\mathrm{var}(\underline{x}_i) < 2\ \mathrm{cov}(\underline{x}_i, \underline{y}_i) \qquad (i = 1, 2, \ldots) \qquad (12.3)$$

or

$$\rho_3 > \frac{1}{2} \frac{\sigma(\underline{x}_i)}{\sigma(\underline{y}_i)} \qquad (12.4)$$

where ρ_3 denotes the positive correlation between $\underline{x}_i$ and $\underline{y}_i$. Hence if there is a strong (positive) correlation between $\underline{x}_i$ and $\underline{y}_j$ then $(\sigma_1^2 - 2c_3)$ may be negative. In the same way we can derive $(\sigma_2^2 - 2c_3)$ is <u>negative or positive</u> depending on the degree of correlation.

(iii) <u>$\sigma_1^2 + c_1 - 2c_3 - 2c_4$.</u> As above we can derive that $(\sigma_1^2 + c_1 - 2c_3 - 2c_4)$ is negative if

$$\frac{\sigma(\underline{x}_i)}{\sigma(\underline{y}_{2i-1})} < 2\,\frac{(\rho_3 + \rho_4)}{(1 + \rho_1)} \qquad (12.5)$$

ρ_4 denoting the undesirable negative correlation between $\underline{x}_{2i}$ and $\underline{y}_{2i-1}$. Hence we conclude that the coefficient $(\sigma_1^2 + c_1 - 2c_3 - 2c_4)$ can be <u>negative</u> or <u>positive</u>. The same conclusion holds for $(\sigma_2^2 + c_2 - 2c_3 - 2c_4)$.

Further we know that σ_1^2 and σ_2^2 are positive. So from Table 10 it follows that there are two possibilities.

(1) Both coefficients a_1 and a_2 are positive. (This always holds for method A.)

(2) One coefficient is positive and one coefficient is negative. It is impossible that both coefficients are negative. (This is not surprising since otherwise the variance would be negative.)

APPENDIX III.13. OPTIMUM COMPUTER TIME ALLOCATION

In this appendix we shall derive the values of M and N, the number of runs of systems 1 and 2, respectively, which minimize the variance of the estimated difference between the responses of systems 1 and 2. As we showed in Table 10 this variance can be approximated by

$$\text{var}(\bar{\underline{d}}) = aM^{-1} + a_2N^{-1} \qquad (13.1)$$

where the particular values of a_1 and a_2 are valid for, say, the case

$$M \geq N \tag{13.2}$$

Further M and N cannot be smaller than N_p, the number of pilot-runs of each system, i.e.,

$$M \geq N_p, \qquad N \geq N_p \tag{13.3}$$

where the constant N_p obviously satisfies

$$0 < N_p < T/(t_1 + t_2) \tag{13.4}$$

since there is a limited amount of computer time available as (13.5) shows.

$$t_1 M + t_2 N = T \qquad (t_1,\ t_2,\ T > 0) \tag{13.5}$$

Hence we have to minimize (13.1) subject to (13.2), (13.3), and (13.5). Now (13.2) and (13.3) together are equivalent to

$$M \geq N \geq N_p \tag{13.6}$$

Condition (13.5) is equivalent to

$$M = t_1^{-1} T - t_1^{-1} t_2 N \tag{13.7}$$

so $M \geq N$ if $N \leq T(t_1 + t_2)^{-1}$. Hence (13.6) and (13.7) together are equivalent to

$$N_p \leq N \leq T(t_1 + t_2)^{-1} \tag{13.8}$$

In this way we have reduced the various restrictions to the single restriction (13.8). Substitution of (13.7) into (13.1) gives

$$\mathrm{var}(\underline{\bar{d}}) = a_1 t_1 (T - t_2 N)^{-1} + a_2 N^{-1} \tag{13.9}$$

To find the value of N that minimizes (13.9) we first consider possibility 1, both coefficients a_1 and a_2 are <u>positive</u>: Determine the first and second derivatives, i.e.,

$$\frac{d[\mathrm{var}(\bar{\underline{d}})]}{dN} = a_1 t_1 t_2 (T - t_2 N)^{-2} - a_2 N^{-2} \tag{13.10}$$

and

$$\frac{d^2[\mathrm{var}(\bar{\underline{d}})]}{dN^2} = 2a_1 t_1 t_2^2 (T - t_2 N)^{-3} + 2a_2 N^{-3} \tag{13.11}$$

Hence the first derivative is zero if N satisfies

$$N = T/\{t_2 \pm (a_1 a_2^{-1} t_1 t_2)^{1/2}\} \tag{13.12}$$

Since

$$T/\{t_2 - (a_1 a_2^{-1} t_1 t_2)^{1/2}\} > T/(t_2 + t_1) \tag{13.13}$$

one solution in (13.12) violates the restriction (13.8). Hence there remains

$$N = N^* = T/\{t_2 + (a_1 a_2^{-1} t_1 t_2)^{1/2}\} \tag{13.14}$$

which indeed gives a minimum since substitution of $N = N^*$ into the second derivative (13.11) yields

$$\left[\frac{d^2[\mathrm{var}(\bar{\underline{d}})]}{dN^2}\right]_{N=N^*} > 0 \tag{13.15}$$

Substitution of N^* into (13.7) yields the corresponding value for M, i.e.,

$$M = M^* = T/\{t_1 + (a_1^{-1} a_2 t_1 t_2)^{1/2}\} \tag{13.16}$$

M^* and N^* are the optimum values unless restriction (13.8) is violated. To find the solution in case of such a violation we

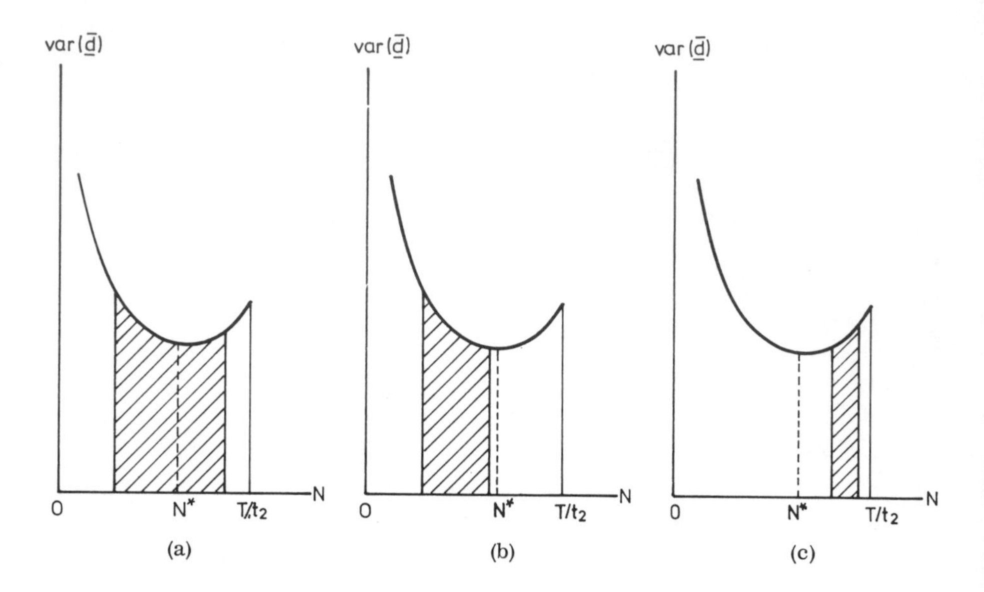

Fig. 13.1. The minimizing value N^* and the (shaded) admissible region for N.

derive the form of the function $\text{var}(\bar{d})$ defined in (13.9). From the formula for the second derivative in (13.11) it follows that

$$\frac{d^2[\text{var}(\bar{d})]}{dN^2} \geq 0 \quad \text{if} \quad 0 \leq N \leq T/t_2 \tag{13.17}$$

From (13.14) it follows that N^* lies within $[0, T/t_2]$ and from (13.4) and (13.8) it follows that the admissible region is part of the interval $[0, T/t_2]$. Hence in the admissible region the second derivative is never negative so $\text{var}(\bar{d})$ decreases monotonically in the interval $0 < N < N^*$ and increases monotonically for $N^* < N < T/t_2$. Therefore we may distinguish the three situations represented in Fig. 13.1 where the shaded area corresponds with the admissible region $[N_p, T/(t_1 + t_2)]$. From Fig. 13.1 N_0 and M_0, i.e., the optimum value of N and via (13.7) the corresponding optimum of M, follow:

(a) If $N_p \leq N^* \leq T/(t_1 + t_2)$ then $N_0 = N^*$ and $M_0 = M^*$.

(b) If $N^* > T/(t_1 + t_2)$ then $N_0 = T/(t_1 + t_2)$ and $M_0 = T/(t_1 + t_2)$.

(c) If $N^* < N_p$ then $N_0 = N_p$ and $M_0 = (T - t_2N_p)/t_1$.

A completely analogous solution can be derived if (13.2) is replaced by $M \leq N$. Next we shall consider possibility 2, one coefficient in (13.9) is positive and the other coefficient is negative. Say that a_1 is negative. From (13.10) it follows that $var(\bar{d})$ decreases as N increases. Therefore we should take N as large as possible. Table 10 and the results of Appendix III.12 show that a negative value of a_1 implies $M \geq N$ (for if $M \leq N$ then a_1 is always positive). The condition $M \geq N$ together with the computer time and the pilot-runs conditions yielded (13.8). So if we want to take N as large as possible and satisfy condition (13.8), then we should take

$$N_0 = T/(t_1 + t_2) \tag{13.18}$$

and therefore

$$M_0 = T/(t_1 + t_2) \tag{13.19}$$

In the same way it can be shown that if a_2 is negative then N should be taken as small as possible. Further a negative value of a_2 implies that $M \leq N$, and this leads again to (13.18) and (13.19).

EXERCISES

1. Prove that in (3.9) $E[(\bar{y} - \mu)^2] = \sigma^2/N + (BIAS)^2$.

2. Prove that $\bar{x}_{SAP}$ defined in (29) is unbiased.

3. Prove that $var(\Sigma_{i=1}^{m} y_i/m) = var(y)\, E(1/m)$.

4. Prove that the estimated coefficients $\hat{\vec{A}} = (\vec{Z}'\vec{Z})^{-1}\vec{Z}'\vec{X}$ in (140) would be unbiased if $\vec{Z}$ were nonstochastic.

5. Suppose we have n independent pairs $(\underline{x}_i, \underline{y}_i)$ $(i = 1, 2, \ldots, n)$. The usual estimator of μ_x/μ_y is $\hat{\underline{\theta}} = \bar{\underline{x}}/\bar{\underline{y}} = \Sigma\, \underline{x}_i / \Sigma\, \underline{y}_i$. Give the jackknife estimator based on forming two groups; see also Rao (1969, Eq. (4)).

6. Prove that the variance of $g(\underline{x})$, defined in (164), is given by

$$\mathrm{var}[g(\underline{x})] = \lambda \exp(-\lambda v)/v - \lambda\xi - \xi^2$$

7. Let $\xi(\lambda, v)$ in (163) be estimated using importance sampling with as new density function the exponential density $h_3(x)$ with modified parameter, say λ'. Prove that the range of the resulting estimator $g_3^*(\underline{x})$ is minimized for $\lambda' = 1/v$ (provided $1/v \leq \lambda$). Prove that the resulting range for $g_3^*(\underline{x})$ is larger than that of $g_1^*(\underline{x})$ and $g_2^*(\underline{x})$ but not larger than that of $g(\underline{x})$.

8. The function $h_0(\vec{R}) = g(\vec{R})\, f(\vec{R})/E(\underline{x})$ in (189) can be a density function if $g(\vec{R}) \geq 0$. What is a second condition; prove that this condition also holds.

9. Prove that when sampling from a symmetric distribution with mean η, the procedure $\underline{y}^* = 2\eta - \underline{y}$ is identical to $\underline{y}^* = F^{-1}(1 - \underline{r})$.

10. Prove that for symmetric distributions the variable $\underline{y}$ and its antithetic partner $\underline{y}^*$ have correlation -1. What is the correlation coefficient when generating antithetic variates from an exponential distribution by means of $\underline{y}^* = F^{-1}(1 - \underline{r})$? What is the correlation between the variate $\underline{x}$ and its linear transformation $\underline{y} = a + b\underline{x}$?

11. Prove that $\underline{r}_2$ is a random number if it is defined as in (226), i.e., $\underline{r}_2 = c - \underline{r}_1$ for $0 \leq \underline{r}_1 \leq c$ and $\underline{r}_2 = 1 + c - \underline{r}_1$ for $c \leq \underline{r}_1 \leq 1$, $\underline{r}_1$ being a random number.

12. Prove that in Fig. 6b $\underline{x}$ and $\underline{y}$ would be completely uncorrelated if we put $\underline{x} = c$ (i.e., $\underline{x}$ is a constant).

13. Consider the simulation of the single-server system for the first-in-first-out and the last-in-first-out queuing discipline. Demonstrate that separate random number generators for the service

and interarrival times do not give exact synchronization. How would you create exact synchronization in this type of problems (also compare job shop simulations)?

14. Simulate the single-server queuing system applying each of the six VRT's and estimate the variance reductions.

15. Apply antithetic variates, common random numbers and their combination in a simple simulation, e.g., an inventory simulation.

NOTES

1. This chapter is based on Kleijnen (1971).

2. VRT's in the estimation of variances are discussed by Beja (1969, pp. 7-8), Fishman (1972, p. 182), Kahn and Marshall (1953, p. 273), Lewis (1972, p. 11) and Matérn (1962); for a general discussion of variance estimation see Raj (1968, pp. 189-199). VRT's in the estimation of the probability distribution itself are examined by Fieller and Hartley (1954) (cf. our section III. 4.4) and by Burt and Garman (1971b), Burt et al. (1970, p. 442), Garman (1972), and Ringer (1965).

3. An exception is Lombaers (1968, p. 253) who studied the decrease of the standard deviation. We also mention the concept of "deficiency" or $1 - \sigma_0^2/\sigma_1^2$ in Andrews et al. (1972, p. 121) where σ_0^2 and σ_1^2 denote the variance of the standard and new estimator, respectively.

4. Some of the VRT's applied in Monte Carlo studies originate from sample survey methods, e.g., stratification and control variates.

5. The "conditional Monte Carlo" technique, applied to stochastic networks, is presented by Burt and Garman (1971b) and Garman (1972). The use of expected values is discussed by Handscomb (1968, pp. 4-5), Kahn and Mann (1957, pp. 15-17), Maxwell (1965, p. 4), and Moy (1965, p. 21). "Virtual" measures of performance are introduced by Carter and Ignall (1970, p. 13; 1972). The statistical estimation method is presented by Kahn and Marshall

(1953, pp. 274-275). Russian roulette and splitting are mentioned by Kahn and Mann (1957), Moy (1965, p. 20), and Tocher (1963, pp. 109-110). Relles (1970) devised a VRT for distribution sampling from Student distributions. Andrews et al. (1972, pp. 55-61, 349-368) discuss VRT's in Monte Carlo experiments, especially experiments with "contaminated" Gaussian distributions. Jessop (1965, p. 161), Bakes et al. (1970, pp. 58-64), and Van Slyke (1963, p. 847) demonstrate that we may try to solve at least part of a problem by analytical methods so that stochastic variation exists only in the remaining part; see also the expected values technique. Multiplex sampling was devised by Evans (1963). A Markov chain method is discussed by Hastings (1970). Easterfield (1961) invented a short cut method for queuing systems with customers arriving in batches. Kosten's technique for independent arrival and service times in queuing systems is presented by De Boer (1969) and Kosten (1948). More VRT's and references are given by Dutton and Starbuck (1971, p. 592), Halton (1970, pp. 7-13), Hammersley and Handscomb (1964), Kahn (1955), Moy (1965, pp. 12-23), Nelson (1966, p. 17), and Wurl (1971, pp. 60-63). Raj (1968) discusses many sampling techniques not in a Monte Carlo context; see also Matérn (1960). In Bauknecht and Nef (1971, pp. 35-153), Fischer, Ragaz, and Stoop improve the efficiency by developing a simulation language (BERSIM) that looks at the queuing systems (especially traffic systems) in a particular way such that fewer random numbers and less computer time are required.

6. The symbol ϵ denotes that $\underline{y}$ belongs to the stratum S_k, i.e., the value of $\underline{y}$ falls within the stratum limits of S_k.

7. The expression after the vertical dash $|$ denotes a condition. So (9) represents the variance, i.e., the expectation of $(x - \mu_k)^2$, under the condition that $\underline{x}$ belongs to stratum k. Remember that $\underline{x}_i$ falls into S_k if $\underline{y}_i$ belongs to S_k.

8. We used equation (5.15) in Cochran (1966, p. 94), but assumed an infinite population; see also Moy (1965, p. 81). Assuming an infinite population is realistic for simulation and Monte Carlo studies, where we can take "infinitely" many observations by taking different random numbers for each observation.

9. Comparing the notation in (6) and (18) demonstrates the advantage of distinguishing between stochastic and nonstochastic variables (by underlining stochastic variables). This convention is not followed by Cochran and Moy. We shall see that Moy overlooked the stochastic character of some variables.

10. We point out that Johnston (1963, p. 276) also defines the MSE, but he mixes up the <u>average</u> value of a sample statistic in, say, N replicated samples and the <u>mean</u> value, i.e., the average over infinitely many replicated samples. Our definition agrees with that in Toro-Vizcarrondo and Wallace (1968, p. 560).

11. After some tedious algebra we derived

$$\begin{aligned} \mathrm{var}(\bar{\underline{w}}_k) &= \mathrm{var}(\bar{\underline{w}}_g)\, P(\underline{n}_k \leq 1) + \mathrm{var}(\bar{\underline{x}}_{kh})\, E(\underline{n}_k \mid \underline{n}_k \geq 2) \\ &\quad + \{E(\bar{\underline{w}}_g) - E(\bar{\underline{w}}_k)\}^2\, P(\underline{n}_k \leq 1) \\ &\quad + \{E(\underline{x}_{kh}) - E(\bar{\underline{w}}_k)\}^2 P(\underline{n}_k \geq 2)\ . \end{aligned}$$

12. We shall return to this type of approach in Section II.4.4; see especially our discussion of (152) in that section.

13. The "quasi-random numbers" method briefly described by Handscomb (1968, p. 10) seems to amount to the same technique as Moy's procedure. Moy's method is related to the stratification variant for the estimation of an integral which as we mentioned at the end of Section III.2.3 is described in Hammersley and Handscomb (1964, p. 55). Recently the Hammersley and Handscomb technique was worked out for simulation by Burt et al. (1970); see also Andréasson (1972b, p. 6), Burt and Garman (1971a, pp. 254-256), Fishman (1972), Garman (1971, p. 11), Shedler and Yang (1971, p. 121). Their procedure resembles antithetic variates (discussed in Section III.6) since k (instead of

two) partner runs are generated and these runs are assumed to be negatively correlated (as in antithetics). We believe, however, that this technique requires more extra programming and running time than antithetics. The reader should consult the article by Burt et al. for additional details.

14. $[a,b)$ denotes the interval from a to b with a included and b excluded.

15. Except for block 1 of Fig. 2 where we use a $\leq$ sign, and Moy uses a $>$ sign, which is probably a printing error.

16. Our first remark is based on a previous publication, see Kleijnen (1969, pp. 294-295).

17. The variables $\underline{y}$ etc. in this section are defined differently from Section III.2. We use the symbols M, q_i, and m where Brenner uses m, p, and n, respectively.

18. As we saw above, in selective sampling each run satisfies $\underline{v}_i = v_i^*$ while runs still differ from each other since the order of events differs.

19. We derived that the two estimators should have a correlation coefficient larger in absolute value than 1/2 under the assumption that both estimators have the same absolute value of their variation coefficient (this coefficient is defined as the standard deviation divided by the mean). This derivation is not given here since its result seems to have only very limited applicability.

20. Moy (1965, p. 47) used a different generalized splitting procedure. His procedure seems less efficient since he estimated a coefficient from only n/J pairs instead of $(n - n/J)$ pairs. Hence his conclusion that splitting into two groups is most efficient, is questionable.

21. The estimator in (123) is in general not unbiased. For in general $E(\underline{x}_1/\underline{x}_2) \neq E(\underline{x}_1)/E(\underline{x}_2)$ or in (123) $E(\hat{\underline{a}}_{k0}) \neq E(\underline{x}\ \underline{z}_k)/E(\underline{z}^2) = a_{k0}$.

22. Johnston (1963, p. 107) assumes that the explanatory variables are fixed numbers, whereas in (124) they are the stochastic variables $\underline{z}_k$. The least squares procedure is completely independent of the stochastic character of $\underline{z}_k$.

23. Beja's result can be easily checked applying the following result given in Scheffé (1964, p. 8): If $\vec{\underline{W}} = \vec{A}\,\vec{\underline{V}}$ then $\vec{\Omega}_w = \vec{A}\,\vec{\Omega}_v\,\vec{A}'$ where $\vec{\Omega}_v$ denotes the covariance matrix of the variables $\underline{v}$, and $\vec{\Omega}_w$ that of $\underline{w}$.

24. An alternative derivation of the least squares character of the estimated optimal coefficients can be found in Moy (1965, pp. 31-33). He does not use the condition $E(\underline{z}_k) = 0$ but instead requires that

$$\bar{\underline{x}} = \sum_{i=1}^{n} \sum_{k=1}^{K} a_k\, \underline{z}_{ki}/n$$

See Moy (1965, p. 33).

25. See Burt et al. (1970), Burt and Garman (1971a, pp. 256-259), Ehrenfeld and Ben-Tuvia (1962, pp. 268-270), Garman (1971, pp. 12-14), Gaver (1969), Gaver and Shedler (1971), Haitsma and Oosterhoff (1964, pp. 31-34), Handscomb (1968, p. 8), Hillier and Lieberman (1968, p. 460-461), Kahn and Marshall (1953, p. 269), Molenaar (1968, pp. 121-122), Newman and Odell (1971, pp. 62-64). Note that Ehrenfeld and Ben-Tuvia's presentation of the control variate estimator is not quite correct since $\bar{W}_1$ in their formula (3.2) should be a constant according to their derivation of the variance of the control variate estimator. Actually $\bar{W}_1$ is estimated in their example.

26. Most authors a priori put the factor a in the control variate estimator (152) equal to 1, so (152) reduces to $\hat{\underline{\mu}}_x = \mu_z(\underline{x} - \underline{z})$.

27. The condition to be imposed upon this control variate estimator in order to achieve an efficiency gain, is given by Maxwell (1965, p. 8). Observe that a minus sign should be placed before the right-hand side of his condition.

28. This section is based on a previous study published in the Dutch language; see Kleijnen (1968b).

29. c is positive since it is the parameter of an exponential distribution.

30. Actually this procedure was applied in (ii), where the original exponential distribution (i.e., the gamma distribution with parameters $p = 1$ and $b = \lambda$) is replaced by the gamma distribution with $p = 2$ and $b = 2/v$.

31. In Kleijnen (1968b, p. 197) it is further proved that even when $\xi(\lambda,v)$ is unknown, we can derive that the variance of $g_1^*(\underline{x})$ is smaller than that of $g(\underline{x})$, but without knowledge of $\xi(\lambda,v)$, we cannot prove that $g_2^*(\underline{x})$ and $g_3^*(\underline{x})$ have a smaller variance than $g(\underline{x})$ or $g_1^*(\underline{x})$. So selecting the approximation that has the smallest variance is impossible.

32. In Kleijnen (1968b, p. 198, line 3) a printing error occurs. The expression for $\mathrm{var}[g_2^*(\underline{x})]$ should be multiplied by the factor 1/2 to correct this error.

33. In Kleijnen (1968b, p. 189) the variances are estimated. These estimates agree very well with the exact values in Table 1.

34. If $\lambda = 0.5$ and $v = 2$ then $h_3^*(x)$ is identical to the old density $f(x)$ so $g_3^*(x) = g(\underline{x})$.

35. The terms pseudorandom and quasirandom have already been used for other stochastic variables.

36. Moy does not consider a stochastic $\underline{m}$ in his derivation of importance sampling. Nevertheless, one of the examples where he applies importance sampling, does have a stochastic number of random numbers.

37. Though Moy does not specify $v_t(r_t)$ exactly as we do in (204) it is easily checked that α_t should be larger than 1 in order to make $v_t(r_t)$ an increasing function of r_t. If $\alpha_t > 1$ then it can further be checked that $v_t(r_t)$ is not negative

for any r_t in $[-\infty,\infty]$. Finally, putting $v_t(r_t)$ equal to 0 for r_t outside $[0,1]$ makes

$$\int_{-\infty}^{\infty} v_t(r_t)\, dr_t$$

equal to 1 so $v_t(r_t)$ is a density function indeed.

38. Compare the following formula for differentiation under the sign of integration, given in Keeping (1962, pp. 392-393): If

$$I(\alpha) = \int_a^b f(x, \alpha)\, dx$$

then

$$\frac{dI(\alpha)}{d\alpha} = \int_a^b \frac{\partial f}{\partial \alpha}\, dx$$

39. More exactly, it is a monotonic nondecreasing function. Nevertheless if we take r_1 and r_2 far enough apart then (222) still holds.

40. E.g., in Appendix I.2 a normal variable $\underline{y}$ was generated from $\Sigma\, \underline{r}_j$. Hence an increase in any r_j leads to an increase in y.

41. Because of the monotonic relation we apply $(1 - r)$ instead of (226). Andréasson (1971a, p. 5) proved that using r and $(1 - r)$ extremizes the (negative) correlation between the input variable $\underline{y}$ and its antithetic partner $\underline{y}_a$. Note that his theorem does not imply that the correlation between the <u>output</u> variables $\underline{x}$ and $\underline{x}_a$ is minimized by the use of r and $1 - r$; compare Andréasson (1971a, p. 25). Recently, Andréasson (1972a) and (1972b) made an interesting study on generalized schemes like $\underline{r}_2 = a \pm \underline{r}$ (mod 1) (and a permutation of the random numbers; see footnote 46). He further investigated the optimal combination of several antithetic techniques belonging to the general scheme. However, such an "optimal" combination (i.e., a combination with minimum variance) may be suboptimal because of extra programming and running time and analysis complications (all runs within a combination are dependent).

42. For the sake of simplicity we take m constant instead of stochastic. This simplification does not affect the argument in (ii).

43. This procedure for generating antithetic random numbers was discovered experimentally by Mr. H. Tilborghs, Katholieke Hogeschool Tilburg.

44. This procedure was suggested to us by Professor H. Lombaers, Technische Hogeschool Delft.

45. Note that in the simulation language GPSS both discrete and continuous stochastic input variables are generated using tables instead of functions (Naylor et al., 1967a, p. 250).

46. Besides switching $\vec{R}_1$ and $\vec{R}_2$ the resulting input variables were adjusted by a factor α; see Moy (1965, p. 71).

47. Moy (1965, p. 76) states that the numbers in his table are the variance ratios $\hat{\sigma}_1^2/\hat{\sigma}_0^2$. However, since he compares these numbers directly with the variance reductions $\{1 - (\hat{\sigma}_1^2/\hat{\sigma}_0^2)\}$ of the other VRT's, we assume that his table actually gives variance reductions.

48. The reader who does not feel familiar with the "power" of a test is referred to Keeping (1962, p. 133). In Chapter V, part B, we shall mention an example where (simultaneous) confidence intervals based on the F-statistic are indeed narrower when applying antithetics since the resulting correlation compensates the loss of degrees of freedom.

49. Section III.8 is based on Kleijnen (1968a).

50. In Kleijnen (1968a, pp. 15, 30-31) we derived the following result. If we apply Tochers "2^K factorial" approach and we assume that this approach works, then var($\underline{\bar{d}}$) is minimal if we apply method A (antithetics only). However, in our discussion of Table 2 in Section III.6 we concluded that Tocher's approach actually does not work.

51. The two variants we simulated are the plans I and II in Table I of Naylor et al. (1967b).

52. We thank Mr.H. Tilborghs (Katholieke Hogeschool Tilburg) who programmed the single-server systems, and Professors T. Wonnacott (Western Ontario University) and D. Graham (Duke University) who assisted with the four-stations systems.

53. Andréasson (1970, p. 19) simulated a multichannel system and found the order C, B, A, D. Tocher (private communication) reported that C gave worse results than B.

54. For method B a more accurate formula for $\text{var}(\underline{\bar{d}})$ could be obtained through addition in Table 10 of the term $\{M^{-2}(M - N)c_1\}$ if $M > N$ or $\{N^{-2}(N - M)c_2\}$ if $M < N$. This term, however, would not allow a simple solution for the optimal values of M and N. For we would have to solve for B:

$$\lambda a_1 M^2 + (2c_1\sigma_2)(-c_1 + \lambda a_2 M^2)^{-1/2} - (\sigma_1^2 + c_1 - 2c_3) = 0$$

where λ is a Lagrangian multiplier. The expression for $\text{var}(\underline{\bar{d}})$ in Table 10 actually represents a pure application of using common random numbers only instead of a mixture where in the last $(M - N)$ or $(N - M)$ runs antithetics are applied; as we shall see in Section III.8.7 such a pure application may be more attractive since its analysis is simpler.

55. We have assumed that we take an equal number (N_p) of observations for each system in the pilot-phase. If we took different numbers of observations then some observations of a system would not be paired with observations of the other system. Boas (1967, p. 291) gives the following formula for the estimation of the covariance using all observations, paired and unpaired ones.

$$\hat{\text{cov}}(\underline{x}, \underline{y}) = \sum_{i=1}^{n} (\underline{x}_i - \bar{\underline{x}})(\underline{y}_i - \bar{\underline{y}})(n_1 n_2) / n(n_1 n_2 - n_1 - n_2 + n)$$

where $(\underline{x}_i, \underline{y}_i)$ are n paired replications, $\bar{\underline{x}}$ is the average of the n_1 ($\geq n$) replications of $\underline{x}$ and $\bar{\underline{y}}$ is the average of the n_2 ($\geq n$) replications of $\underline{y}$; all $\underline{x}$ are mutually independent, all $\underline{y}$ are mutually independent.

56. In systems with a stochastic number of random numbers per run $\underline{T}_p$ will indeed be stochastic. In the single-server system where we decide to simulate a fixed number of customers, $\underline{T}_p$ is nonstochastic.

57. If $\underline{x} = g(\underline{y})$ then in general $E[\underline{x}] \neq g(E[\underline{y}])$ unless $g(\underline{y})$ is a linear function.

58. Ghurye and Robbins (1954) investigated the estimation of the difference between two means when the sample observations are independent. In the formula for the optimal sample sizes they replaced the unknown variances by the estimators based on a pilot-sample. They derived that in the resulting two-stage procedure the variance of $\bar{\underline{d}}$ asymptotically approaches the variance of $\bar{\underline{d}}$ for known variances σ_1^2 and σ_2^2.

59. These results were obtained by Mr. H. Tilborghs (Katholieke Hogeschool Tilburg).

60. For the sake of convenience we suppose that M is even. Further we let M denote the optimum number of runs resulting from Table 11 and (277) through (280). Observe that in (292) we want to estimate the variance of the average $\bar{\underline{x}}$ while in (288) through (290) we estimated the variance of an individual observation $\underline{x}$.

61. In (295) we first apply the central limit theorem to the averages of the antithetic pairs; this explains why $\bar{\underline{x}}_1$ is asymptotically normal. The central limit theorem also implies that $\bar{\underline{x}}_2$ is asymptotically normal. Hence $\bar{\underline{x}}$ is asymptotically normal.

62. Note that $\bar{\underline{x}}$ and $\bar{\underline{y}}$ are dependent even if we apply method A after the pilot-phase since in the pilot-phase $\underline{x}_i$ and $\underline{y}_i$ ($i = 1, \ldots, N_p$) are dependent. Observe further that normal marginal distributions do not necessarily imply a normal bivariate distribution; see Ruymgaart (1973).

63. Fishman (1967, p. 6) states that the "converse problem of minimizing V for a given T has the same solution". Comparing (305) and (306) with (275) and (276) shows that both solutions look alike but nevertheless are not identical.

64. A different approach (not in a simulation context) is followed by Sedransk (1971).

65. We owe this proof to Professor M. Euwe (Katholieke Hogeschool Tilburg).

REFERENCES

Andréasson, I.J. (1970). The Application of Two Variance-Reducing Techniques on a Queueing Simulation in Simula-67, Report NA 70.21, Department of Information Processing, The Royal Institute of Technology, Stockholm.

Andréasson, I.J. (1971a). On the Generation of Negatively Correlated Random Numbers, Report NA 71.32, Department of Information Processing, The Royal Institute of Technology, Stockholm.

Andréasson, I.J. (1971b). Antithetic and Control Variate Methods for the Estimation of Probabilities in Simulations, Report NA 71.41, Department of Information Processing, The Royal Institute of Technology, Stockholm.

Andréasson, I.J. (1972a). Combinations of Antithetic Methods in Simulation, in Working Papers, Vol. 1, Symposium Computer simulation versus analytical solutions for business and economic models, Graduate School of Business Administration, Gothenburg, Sweden.

Andréasson, I.J. (1972b). Antithetic Methods in Queuing Simulations, Report NA 72.58, Department of Computer Science, Royal Institute of Technology, Stockholm.

Andrews, D.F., P.J. Bickel, F.R. Hampel, P.J. Huber, W.H. Rogers, and J.W. Tukey (1972). Robust Estimation of Location, Princeton University Press, Princeton, N.J.

Arvesen, J.N. and D.S. Salsburg (1975). "Approximate tests and confidence intervals using the jackknife," in Perspectives in Biometrics (R. Elashoff, ed.), Academic, New York.

Bakes, M.D., M.J. Bramson, S. Freckleton, P.C. Roberts, and D. Ryan (1970). "Stochastic-network reduction and sensitivity techniques in a cost effectiveness study of a military communications system," Operational Res. Quart., 21, Spec. Conf. Iss., 45-67.

Bauknecht, K. and W. Nef (eds.) (1971). Digitale Simulation (Digital Simulation), Springer, Berlin.

Beja, A. (ca. 1969). Multiple Control Variates in Monte Carlo Simulation (with Application to Queuing Systems with Priorities), W. P. No. 80/70, The Leon Recanati Graduate School of Business Administration, Tel-Aviv University, Tel-Aviv. (Also published in: Developments in Operations Research, Vol. 1. B. Avi-Itzhak, ed.), Gordon and Breach, New York.

Boas, J. (1967). "A note on the estimation of the covariance between two random variables using extra information on the separate variables," Stat. Neerl., 21, 291-292.

Brenner, M.E. (1963). "Selective sampling--a technique for reducing sample size in simulation of decision making problems," J. Ind. Eng., 14, 291-296. (Also published in: Computer Simulation of Human Behavior, by J. M. Dutton and W.H. Starbuck, Wiley, New York.)

Burt, J. M. and M.B. Garman (1971a). "Monte Carlo techniques for stochastic PERT network analysis," Infor, 9, 248-262.

Burt, J.M. and M.B. Garman (1971b). "Conditional Monte Carlo: a simulation technique for stochastic network analysis," Management Sci., 18, 207-217.

Burt, J.M., D.P. Gaver, and M. Perlas (1970). "Simple stochastic networks: some problems and procedures," Naval Res. Logistics Quart., 17, 439-459.

Carter, G. and E. Ignall (1970). A Simulation Model of Fire Department Operations: Design and Preliminary Results, R-632-NYC, The New York City Rand Institute, New York. (Also published in IEEE Trans. System Sci. Cybernetics, Oct. 1970.)

Carter, G. and E. Ignall (1972). Virtual Measures for Computer Simulation Experiments, P-4817. The Rand Corp., Santa Monica, Calif.

Clark, C.E. (1959). The Utility of Statistics of Random Numbers, System Development Corp., Santa Monica, Calif.

Clark, C.E. (1961). "Importance sampling in Monte Carlo analyses," Operations Res., 9, 603-620. (Also published in: Computer Simulation of Human Behavior, by J. M. Dutton and W.H. Starbuck, Wiley, New York.)

Cochran, W.G. (1966). Sampling Techniques, 2nd ed., Wiley, New York.

Conway, R.W. (1963). "Some tactical problems in digital simulation," Management Sci. 10, 47-61.

Conway, R.W., B.M. Johnson, and W.L. Maxwell (1959). "Some problems of digital systems simulation," Management Sci., 6, 92-110.

De Boer, J. (1969). "Toepassing van Kosten's lotingsas in simulaties" (Applying Kosten's sampling axis to simulation), Stat. Neerl., 23, 243-248.

Dutton, J.M. and W.H. Starbuck (1971). Computer Simulation of Human Behavior, Wiley, New York.

Easterfield, T.E. (1961). "A short cut in a class of simulation problems," Operational Res. Quart., 12, 221-225.

Ehrenfeld, S. and S. Ben-Tuvia (1962). "The efficiency of statistical simulation procedures," Technometrics, 4, 257-275.

Emshoff, J.R. and R.L. Sisson (1971). Design and Use of Computer Simulation Models, Macmillan, New York.

Evans, D.H. (1963). "Applied multiplex sampling," Technometrics, 5, 341-359.

Farmer, J.A. (1966). Statistical Aspects of Digital Simulation, The Thomson Organisation, paper read at the symposium on "Simulation techniques and languages," Brunel College, Department of Mathematics, London.

Fieller, E.C. and H.O. Hartley (1954). "Sampling with control variables," Biometrika, 41, 494-501.

Fishman, G.S. (1967). Digital Computer Simulation: The Allocation of Computer Time in Comparing Simulation Experiments, RM-5288-1-PR, The Rand Corp., Santa Monica, Calif. (Also published in Operations Res., 16, 1968, 280-295. Erratum in Operations Res., 16, 1968, 1087.)

Fishman, G.S. (1972). "Variance reduction in simulation studies," Stat. Computation Simulation, 1, 173-182.

Fishman, G.S. and P.J. Kiviat (1967). Digital Computer Simulation: Statistical Considerations, RM-5387-PR, The Rand Corp., Santa Monica, Calif.

Fisz, M. (1967). Probability Theory and Mathematical Statistics, 3rd ed., Wiley, New York.

Fraser, D.A.S. (1957). Nonparametric Methods in Statistics, Wiley New York.

Garman, M.B. (1971). Variance Reduction in Large-Scale Computer Simulations, Working paper No. CP-338, Center for Research in Management Science, University of California, Berkeley, Calif.

Garman, M.B. (1972). "More on conditioned sampling in the simulation of stochastic networks," Management Sci., 19, 1972, 90-95.

Garman, M.B. (1973). On the "Variate-Matching" Problem of Digital Simulation, Working paper no. 350, Center for Research in Management Science, University of California, Berkeley, Calif.

Gaver, D.P. (1969). Statistical Methods for Improving Simulation Efficiency, Carnegie-Mellon University, Pittsburgh, Pennsylvania, 1969. (Also in: Proc. 3rd Ann. Conf. Applications Simulation, Los Angeles, Dec. 1969.)

Gaver, D.P. and G.S. Shedler (1971). "Control variable methods in the simulation of a model of a multiprogrammed computer system," Naval Res. Logistics Quart., 18, 435-450.

Ghurye, S.G. and H. Robbins (1954). "Two-stage procedures for estimating the difference between means," Biometrika, 41, parts 1 and 2, 146-152.

Goldberger, A.S. (1964). Econometric Theory, Wiley, New York.

Goodman, A.S., P.A.W. Lewis, and H.E. Robbins (1973). "Simultaneous estimation of large numbers of extreme quantiles in simulation experiments," Commun. Stat.

Gray, H.L. and W.R. Schucany (1972). The Generalized Jackknife Statistic, Marcel Dekker, New York.

Gürtler, H. (1969). Kwantitative Modelle zur Optimierung des Schalterverkehrs in Einem Postamt (Quantitative models for optimizing traffic at counters in a post office), Doctoral dissertation, Wilhelms-Universität, Münster (Germany).

Haitsma, A.H. and J. Oosterhoff (1964). Monte Carlo Methoden (Monte Carlo methods), Rapport S 265 (C13), Leergang Besliskunde, Hoofdstuk XVI, Stichting Mathematisch Centrum, Amsterdam.

Halton, J.H. (1970). "A retrospective and prospective survey of the Monte Carlo method," Siam Rev., 12, 1-63.

Hammersley, J.M. and D.C. Handscomb (1964). Monte Carlo Methods, Wiley, New York; Methuen, London.

Hammersley, J.M. and J. Mauldon (1956). "General principles of antithetic variates," Proc. Cambridge Phil. Soc., 52, 476-481.

Hammersley, J.M. and K.W. Morton (1956). "A new Monte Carlo technique: antithetic variates," Proc. Cambridge Phil. Soc., 52, 449-475.

Handscomb, D.C. (1968). Variance Reduction Techniques, Presented at the Symposium on the design of computer simulation experiments, Duke University, Durham, N.C., 14-16 Oct. 1968. (Also published in The Design of Computer Simulation Experiments, T.H. Naylor, ed., Duke University Press, Durham, N.C., 1969.)

Harling, J. (1958). "Simulations techniques in operational research," Operational Res. Quart., 9, 9-21.

Hastings, W.K. (1970). "Monte Carlo sampling methods using Markov chains and their applications," Biometrika, 57, 97-109.

Henzler, H. (1970). Die Optimierung der Schichtenbildung beim Zufallsgesteuerten Geschichteten Stichprobenverfahren (The optimalisation of strata formation in random stratified sampling), Doctoral dissertation, Ludwig-Maximilian University, Munich.

Hillier, F.S. and G.J. Lieberman (1968). Introduction to Operations Research, Holden-Day, San Francisco, Calif., Chap. 14.

Ignall, E.J. (1972). "On experimental designs for computer simulation experiments," Management Sci., 18, 384-388.

Jessop, W.N. (1956). "Monte Carlo methods and industrial problems," Appl. Stat., 5, 158-165.

Johnston, J. (1963). Econometric Methods, McGraw-Hill, New York.

Kahn, H. and I. Mann (1957). Monte Carlo, Report P-1165, The Rand Corp., Santa Monica, Calif.

Kahn, H. (1955). Use of Different Monte Carlo Sampling Techniques, Report P-766, The Rand Corp., Santa Monica, Calif.

Kahn, H. and A.W. Marshall (1953). "Methods of reducing sample size in Monte Carlo computations," J. Operations Res. Soc. Amer., 1, 263-278.

Keeping, E.S. (1962). Introduction to Statistical Inference, Van Nostrand, Princeton, N.J.

King, G.W. (1953). "The Monte Carlo method as a natural mode of expression in operations research," J. Operations Res. Soc. Amer., 1, 46-51.

Kleijnen, J.P. (1968a). Increasing the Reliability of Estimates in the Simulation of Systems: Negative and Positive Correlation Between

Runs, Preliminary report, Los Angeles, April 1968(a). (Also presented at the European meeting on statistics, econometrics and management science of IMS, TIMS, ES and IASPS, Amsterdam, 2-7 Sept. 1968. Obtainable from the author at the Katholieke Hogeschool, Tilburg, Netherlands.)

Kleijnen, J.P.C. (1968b). "Een toepassing van 'importance sampling'" (An application of importance sampling), Stat. Neerl., 22, 179-198. (Also published as report EIT, no. 2, of the Economisch Instituut Tilburg, Econometrische Afdeling, Tilburg, Netherlands.)

Kleijnen, J.P. (1969). "Monte Carlo techniques: a comment," in The Design of Computer Simulation Experiments (T.H. Naylor, ed.), Duke University Press, Durham, N.C.

Kleijnen, J.P.C. (1971). Variance Reduction Techniques in Simulation, Doctoral dissertation, Katholieke Hogeschool, Tilburg (Netherlands).

Kosten, L. (1948). "On the measurement of congestion quantities by means of fictitious traffic," Het PTT Bedrijf, 2, 15-25.

Kosten, L. (1968). Statistische Aspecten van Simulatie (Statistical aspects of simulation), Mimeographed notes, Afdeling Algemene Wetenschappen, Technische Hogeschool, Delft (Netherland).

Lewis, P.A.W. (1972). Large-Scale Computer-Aided Statistical Mathematics, Naval Postgraduate School, Monterey, Calif. (In Proc. Computer Sci. Stat., 6th Ann. Symp. Interface, Western Periodical Co., Hollywood, Calif., 1973.)

Lombaers, H.J.M. (1968). "Enige statistische aspecten van simulatie" (Some statistical aspects of simulation), Stat. Neerl., 22, 249-255.

Marshall, A.W. (1958). Experimentation by simulation and Monte Carlo, Report P-1174, The Rand Corp., Santa Monica, Calif.

Matérn, B. (1962). "On the use of information on supplementary variables in estimating a distribution," Rev. Int. Stat. Inst., 30, 121-135.

Matérn, B. (1970). Spatial Variation. Meddelanden fran statens skogsforskningsinstitut, band 49, nr. 5, 1960. (Reproduced by offset, Department of Forest Biometry, Royal College of Forestry, Stockholm, August 1970.)

Maxwell, W.L. (1965). Variance Reduction Techniques, The Rand Corp., Santa Monica, Calif.

Mayne, D.Q. (1966). "A gradient method for determining optimal control of nonlinear stochastic systems," Proc. 2nd IFAC Symp. Theory Self-Adaptive Control Systems, P.H. Hammond, ed., Plenum Press, New York.

Meier, R.C., W.T. Newell, and H.L. Pazer (1969). Simulation in Business and Economics, Prentice-Hall, Englewood Cliffs, N.J.

Mickey, M.R. (1959). "Some finite population unbiased ratio and regression estimators," J. Amer. Stat. Assoc., 54, 594-612.

Mihram, G.A. (1972). Simulation: Statistical Foundations and Methodology, Academic, New York.

Mize, J.H. (1973). Multiple Sequence Random Number Generators, School of Industrial Engineering and Management, Oklahoma State University, Stillwater, Oklahoma. Paper prepared for 1973 Winter Simulation Conference, San Francisco, Calif.

Molenaar, W. (1968). "Simulatie" (Simulation), in: Minimaxmethode, Netwerk Planning, Simulatie (J. Kriens, F. Gobel, and W. Molenaar, eds.) Leergang Besliskunde, I.8, Mathematisch Centrum, Amsterdam.

Mosteller, F. and J.W. Tukey (1968). "Data analysis, including statistics," in The Handbook of Social Psychology, 2nd ed., Vol. 2, (G. Lindzey and E. Aronson, eds.), Addison-Wesley, Reading, Pa.

Moy, W.A. (1965). Sampling Techniques for Increasing the Efficiency of Simulations of Queuing Systems, Ph.D. dissertation, Industrial engineering and management science, Northwestern University, August 1965. (Obtainable from University Microfilms, Inc., Ann Arbor, Michigan, order 66-2730.)

Moy, W.A. (1966). Practical Variance-Reducing Procedures for Monte Carlo Simulations, University of Wisconsin, 7 Oct. 1966. (Also published in: The Design of Computer Simulation Experiments, T.H. Naylor, ed., Duke University Press, Durham, N.C., 1969 and in Computer Simulation Experiments with Models of Economic Systems, by T.H. Naylor, Wiley, New York, 1971.)

Naylor, T.H. (1971). Computer Simulation Experiments with Models of Economic Systems, Wiley, New York.

Naylor, T.H., J.L. Balintfy, D.S. Burdick, and K. Chu (1967a). Computer Simulation Techniques, Wiley, New York.

Naylor, T.H., K. Wertz, and T.H. Wonnacott (1967b). "Methods for analyzing data from computer simulation experiments," Commun. ACM, 10, 703-710.

Nelson, R.T. (ca. 1966). Systems, Models and Simulation, Mimeographed notes, Graduate School of Business Administration, University of California, Los Angeles, Calif.

Newman, T.G. and P.L. Odell (1971). The Generation of Random Variates, Griffin, London.

Orcutt, G.H., M. Greenberger, J. Korbel and A.M. Rivlin (1961). Microanalysis of Socioeconomic Systems: A Simulation Study. Harper and Row, New York.

Page, E.S. (1965). "On Monte Carlo methods in congestion problems: II: Simulation of queuing systems," Operations Res., 13, 300-305.

Price, C.M. (1972). The Design and Implementation of Generalised Simulation Models for the Mining Industry, in Working papers, Vol. 2, Symposium Computer simulation versus analytical solutions for business and economic models, Graduate School of Business Administration, Gothenburg (Sweden).

Pugh, E.L. (1966). "A gradient technique of adaptive Monte Carlo," SIAM Rev., 8, 346-355.

Quenouille, M. (1956). "Notes on bias in esimation," Biometrika, 43, 353-360.

Radema, F.W. (1969). Versnellingstechnieken bij Systeemsimulatie (Variance reduction techniques in systems simulation), Groep Bedrijfseconometrie, Technische Hogeschool, Eindhoven (Netherlands).

Raj, D. (1968). Sampling Theory. McGraw-Hill, New York.

Rao, J.N.K. (1969). "Ratio and regression estimators," in New Developments in Survey Sampling, N.L. Johnson and H. Smith, eds., Wiley-Interscience, New York.

Relles, D.A. (1970). "Variance reduction techniques for Monte Carlo sampling from Student distribution," Technometrics, 12, 499-515.

Ringer, L.J. (ca. 1965). Simulation of PERT Project Completion Times by Stratified Sampling Methods, Technical Report no. 3, Texas A and M Research Foundation, processed for Defense Documentation Center, Defense Supply Agency. (Obtainable from Clearinghouse for Federal Scientific and Technical Information.)

Ruymgaart, F.H. (1973). "Non-normal bivariate densities with normal marginals and linear regression functions," Stat. Neerl., 27, 11-17.

Satterthwaite, F.E. (1946). "An approximate distribution of estimates of variance components," Biometrics, 2, 110-114.

Scheffé, H. (1964). The Analysis of Variance, Wiley, New York.

Sedransk, J. (1971). "Precision specifications for the estimation of treatment differences," Biometrics, 27, 673-680.

Shedler, G.S. and S.C. Yang (1971). "Simulation of a model of paging system performance," IBM Systems J., 10, 113-128.

Tocher, K.D. (1963). The Art of Simulation, The English Universities Press, London.

Toro-Vizcarrondo, C. and T.D. Wallace (1968). "A test of the mean square error criterion for restrictions in linear regression," J. Amer. Stat. Assoc., 63, 558-572.

Tukey, J.W. (1958). "Bias and confidence in not quite large samples" (abstract), Ann. Math. Stat., 29, 614.

Van Slyke, R.M. (1963). "Monte Carlo methods and the PERT problem," Operations Res., 11, 839-860.

Williams, W.H. (1964). "Sample selection and the choice of estimator in two-way stratified populations," J. Amer. Stat. Assoc., 59, 1054-1062.

Wurl, H. (1971). Die Anwendung der Simulationstechnik zur Betriebswirtschaftlichen Beurteilung Industrieller Projekte in Entwicklungslandern (The application of the simulation technique for the economic evaluation of industrial projects in underdeveloped countries), Duncker and Humblot, Berlin.